CONSTRUCTION CONTRACT LAW

CONSTRUCTION CONTRACT LAW

JOHN J.P. KROL, ESQ.

JOHN WILEY & SONS, INC.
New York • Chichester • Brisbane • Toronto • Singapore

This text is printed on acid-free paper.

Library of Congress Cataloging in Publication Data:

Krol, John J. P., 1951–
Construction contract law / by John J.P. Krol.
p. cm. — (Wiley series of practical construction guides)
Includes index.
ISBN 0-471-57414-7 (cloth : alk. paper)
1. Construction contracts—United States. I. Title. II. Series.
KF902.K76 1993
343.73'07869—dc20
[347.3037869] 92-38225
CIP

10 9 8 7 6 5 4 3 2 1

WILEY SERIES OF PRACTICAL CONSTRUCTION GUIDES

M.D. Morris, P.E., Series Editor

Marvey V. Debo and Leo Diamant
CONSTRUCTION INSPECTION: A FIELD GUIDE TO PRACTICE, Second Edition

Cortland A. Collier and Don A. Halperin
CONSTRUCTION FUNDING: WHERE THE MONEY COMES FROM, Second Edition

Walter Podolny and John B. Scalzi
CONSTRUCTION OF CABLE-STAYED BRIDGES, Second Edition

Edward J. Monahan
CONSTRUCTION OF AND ON COMPACTED FILLS

Ben C. Gerwick, Jr.
CONSTRUCTION OF OFFSHORE STRUCTURES

Leo Diamant
CONSTRUCTION ESTIMATING FOR GENERAL CONTRACTORS

Richard G. Ahlvin and Vernon Allen Smoots
CONSTRUCTION GUIDE FOR SOILS AND FOUNDATIONS, Second Edition

B. Austin Barry
CONSTRUCTION MEASUREMENTS, Second Edition

John E. Traister and Paul Rosenberg
CONSTRUCTION ELECTRICAL CONTRACTING, Second Edition

Harold J. Rosen and Tom Heineman
CONSTRUCTION SPECIFICATIONS WRITING: PRINCIPLES AND PROCEDURES, Third Edition

Leo Diamant and C.R. Tumblin
CONSTRUCTION COST ESTIMATES, Second Edition

Thomas C. Schleifer
CONSTRUCTION CONTRACTORS' SURVIVAL GUIDE

A. C. Houlsby
CONSTRUCTION AND DESIGN OF CEMENT GROUTING: A GUIDE TO GROUTING IN ROCK FOUNDATIONS

David A. Day and Neal B. H. Benjamin
CONSTRUCTION EQUIPMENT GUIDE, Second Edition

Terry T. McFadden and F. Lawrence Bennett
CONSTRUCTION IN COLD REGIONS: A GUIDE FOR PLANNERS, ENGINEERS, CONTRACTORS, AND MANAGERS

S. Peter Volpe and Peter J. Volpe
CONSTRUCTION BUSINESS MANAGEMENT

Julian R. Panek and John Philip Cook
CONSTRUCTION SEALANTS AND ADHESIVES, Third Edition

J. Patrick Powers
CONSTRUCTION DEWATERING: NEW METHODS AND APPLICATIONS, Second Edition

William R. Park and Wayne B. Chapin
CONSTRUCTION BIDDING: STRATEGIC PRICING FOR PROFIT, Second Edition

Ellis J. Krinitzsky, James P. Gould, and Peter H. Edinger
FUNDAMENTALS OF EARTHQUAKE RESISTANT CONSTRUCTION

Ben C. Gerwick, Jr.
CONSTRUCTION OF PRESTRESSED CONCRETE STRUCTURES, Second Edition

John J.P. Krol
CONSTRUCTION CONTRACT LAW

PREFACE

The difficulty in writing this book lies with its title *Construction Contract Law*. The fact is there is no one construction contract law. We operate in a federal union, composed of 50 different states, with, on some issues, as many as 50 different opinions on how to deal with contracts. Add to this the overlay of a federal system with its own idiosyncracies when dealing with federal construction contracts, and you have some idea of the magnitude of my problem.

The cases used here are for illustration only. In using them, I did not attempt to demonstrate the current statutory or case law of any particular jurisdiction. What I have attempted to do is provide you, the contractor and layman, with some insight into the judicial mind, its thought processes, and the very real result of those processes. I have done this in the hope that it will better enable you, first, to recognize your own problems in these cases, and second, to empower you to be more fully prepared to discuss these with your own attorney. Your own attorney is necessary. He or she is necessary to lead you through the labyrinth that your particular state has laid, by statute or by prior court decisions, upon the general principles and considerations contained in this volume. Your counsel will know enough to consult the particular case law of your jurisdiction.

In our world, which threatens to leave order behind, this book resurrects the ghosts of the past. The jurists of the past have striven, by their words, to impart widom, to do justice. Why? Because justice is all there is in a democracy. Justice is that plea, of one citizen, that will be answered. Justice means that the plea of that one citizen is the equal to that of any other citizen, or any corporation, or of any multinational corporation. That single, sole plea is evaluated on its merits and on its merits alone. This is justice.

The volume cannot provide you with the current law of your particular jurisdiction. What it can provide is some idea of the breadth and range of the intellectual and moral underpinnings of that system; a system by which justice is surely done.

New York
September 1992

JOHN J.P. KROL

NOTE

The author thanks Kenneth Lazaruk, Esq. of New York City and John Moss Hinchcliff, Esq. of Ithaca for their efforts in reviewing this text.

ABOUT THE AUTHOR

John J.P. Krol was born in New York City. He received his engineering degree from Manhattan College and his law degree from Fordham University. He is a licensed Professional Engineer in New York State, and is admitted to the New York State courts as well as to the Eastern, Southern, and Northern District Federal courts in New York. He is a member of the bar of the United States Court of Appeals for the Federal Circuit, the United States Court of Federal Claims, and the United States Supreme Court. He is a member of the panel of arbitrators of the American Arbitration Association, of the New York State Bar Association, and of the American Society of Civil Engineers.

He is in private practice in Rockville Centre and New York City.

To Lizanne, for all.

TABLE OF CONTENTS

CHAPTER 1

FORMATION OF CONTRACTS

This book will alert you, the contractor, to the pitfalls contained in any construction contract. On one hand, you will need merely to be aware of a particular provision and its consequences. On the other, there are times when the clear and present danger of a contractual provision cries out to you to consult your attorney. If smart is accepting only half of any advice offered, then brilliant is knowing which half to accept.

As an attorney who has read many contracts and advised many contractor clients, I am often shocked to discover that those provisions that most alarm me and my fellow attorneys were hardly noticed by my clients. *Prior notice* provisions, no damages for delay clauses, and conditional payment provisions are among the items that will be discussed here. Too frequently they come to the contractor's attention when the contractor comes to his attorney. The premise of this book is that I will act as your "Dutch uncle," whispering when to accept a contractual provision and move on, and when to go directly to your telephone and call your attorney. Of course, any book can only offer generalities. The individual nature of any particular contract will always dominate the legal evaluation of its effect. But this book can help you know what you don't know and can provide a solid basis for an intelligent discussion with the attorney you select.

1.1 FORMAL REQUISITES

§ 1.1.1 Delivery

If there were a title that would put you off, "Formal requisites" may be it. Still, the law does require that certain minimal standards be met before it will say that a contract has come into being. Most construction contracts come in standardized guises. The American Institute of Architects provides a popular form of construction contract. Large municipalities and state and interstate agencies all have their own forms of contract and their own peculiar boilerplate. Federal contracts could be, and are, the subject of many erudite volumes. What I intend to do is distill from all those sources, the essentials of the construction contract. To do that, we must begin at the beginning, at those formal requisites.

The law provides a battery of requisites, but many of them need not concern us here. Three perennial concerns are delivery, signature, and content.

Signed, sealed, and delivered. We almost nod our heads in agreement, but what does it mean? Return to, perhaps, your first civics class. In *Marbury v. Madison*, 1 Cranch 137, 2 L.Ed. 60 (1803), outgoing President John Adams had signed the commissions of a number of justices of the peace for the District of Columbia. Onto those commissions had been affixed the seal of the United States. But the appointments were never delivered to those prospective justices.

The Supreme Court held that those prospective justices had a right to those commissions. Without delivery, their commissions never were effective. The Court ordered delivery of the commissions. This is a precise analogy to the delivery requirement in connection with a contract.

When a contractor approaches his attorney with a contract problem, it is rarely a delivery problem. In fact, the courts have held that when there exists sufficient evidence exhibiting the contracting party's intent to assume both the benefits and the burden of a contract, then physical delivery itself is unnecessary. *Birch v. McNall*, 19 A.D.2d 850, 244 N.Y.S.2d 60 (4th Dept. 1963).

In the construction industry, the issue of delivery of a contract most frequently arises where a party to the purported contract contests its very existence. Such a case would arise where one party expects performance, and performance is refused. This is not typically the situation where the contractor supplies significant amounts of labor and materials only to find that the owner refuses to pay. In that instance, the court would find that the conduct of the parties indicated an existing contract.

Of course, we cannot forget the moral of *Marbury v. Madison*. In general, a contract is not effective without delivery of the contract; but delivery need not be absolute. A contract may be delivered conditionally; that is, not to become effective without the occurrence of some event or condition. In such a situation, the contract is not effective until the occurrence of the event or the fulfillment of the condition.

The law affords such weight to the conduct of the parties that it will presume the delivery of a contract where the acts of the parties reflect their recognition of contractual obligations. *Sarasohn v. Kamaiky*, 193 N.Y. 203, reh'g den. 194 N.Y. 535 (1908).

§ 1.1.2 Signature

Like delivery, the effect of a signature on a contract is dependent on the circumstances. Parties indicating an intent to be bound by their actions or otherwise will be held to an unsigned contract. There is such a creature as an oral contract. However, the problem with proof of an oral contract gives credence to the old saw: "An oral contract isn't worth the paper it's printed on." A person agreeing to the terms of a contract and governing his or her actions pursuant to those terms will be bound by that contract even if he or she has not signed it. It is unnecessary that a written agreement be signed by a party, in order to bind that party. *Soundview Woods, Inc. v. Mamaroneck*, 14 Misc.2d 866, 178 N.Y.S.2d 800 (West. Co. 1958), aff'd 9 A.D.2d 789, 193 N.Y.S.2d 1021 (2nd Dept. 1959).

Signature by initials has been found to be valid, as will signature by lead pencil. In a case involving a dispute between a general contractor on a large construction contract and his or her subcontractor, a provision providing that the subcontractor would be paid $30 per cubic yard for rock excavation in trenches and footings, and $20 per cubic yard for mass rock excavation, was inked in with pen on an otherwise printed contract. A provision handwritten into a printed contract, when initialed in the margin by both parties, is binding on both parties. *Application of Vincent J. Smith, Inc.*, 19 A.D.2d 763, 241 N.Y.S.2d 507 (3rd Dept. 1963).

This almost casual consideration is deceiving, however. In *Aluminum Fare, Inc. v. Abdella*, 90 A.D.2d 603, 456 N.Y.S.2d 184 (3rd Dept. 1982); app. den. 58 N.Y.2d 606, 460 N.Y.S.2d 1025 (1983), the court interpreted an agreement in connection with the construction of a swimming pool. The court found that no agreement existed between the owner and the contractor. The contractor's written proposal was never signed, and the owner and contractor subsequently, and orally, altered a number of terms. The price was increased as was the pool size. There was never a "meeting of the minds," which creates an enforceable contract. The contractor was awarded the reasonable value of his service, labor, and materials.

§ 1.1.3 Contents

The general rule is that a person signing a contract is bound by that contract whether or not he or she has read it. Were the rule otherwise, a written agreement would be worthless, for the question of what the parties had agreed to would be left to the vagaries of conflicting oral testimony.

This rule especially must be kept in mind when dealing with the so-called boilerplate, the pages upon pages of general provisions found in many standardized contracts. The owner will hold you to those provisions. It is just those provisions that the contractor does *not* review when preparing and submitting his or her bid. The unfortunate contractor first becomes aware of those provisions, and the consequences of those provisions, when he visits his attorney shortly after the contractual relationship has broken down.

In *Business Intelligence Services, Inc. v. Hudson*, 580 F.Supp. 1068 (D.C. N.Y. 1984), a former senior consultant was held to the terms of a noncompetition clause. The clause was contained in an employment contract, which the consultant had failed to read.

The courts have stated bluntly that if a party to a contract could have read that contract in its entity, failure to read it is gross negligence. If the party could not read the contract, then failure to find someone who could read it for him or her is equally negligent. The court found that in either case, the agreement is binding upon the contracting party. *Gillman v. Chase Manhattan Bank, N.A.*, 135 A.D.2d 448, 521 N.Y.S.2d 729 (2nd Dept. 1987), appeal granted 72 N.Y.2d 803, 532 N.Y.S.2d 369 (1988), aff'd 73 N.Y.2d 1, 537 N.Y.S.2d 787 (1988).

In many states, parties need not agree to every aspect of the agreement before an enforceable contract is reached. The law no longer holds that a dispute over material terms will prevent the formation of a binding contract. The law does hold that when the particular dispute over material terms indicates a lack of intention to contract, then no contract results. *J. Baranello & Sons v. Hausmann Industries, Inc.*, 571 F.Supp. 333 (D.C. N.Y. 1983).

1.2 CERTAINTY

§ 1.2.1 Rules of Construction

As a general rule, the law will find that a contract is sufficiently definite and certain when it contains matter that enables a court to ascertain the terms and conditions governing the conduct of the parties. A contract can be made certain by reference to extrinsic matters, an extrinsic document or transaction that can clarify the contract and make it enforceable. The courts do look for three minimum requirements: price, quantity, and time.

§ 1.2.2 Price

In every contract, price is an essential ingredient. A contract is enforceable by the court if the parties to the contract provide the court a practical method to determine the price. However, if the determination of the price by additional formulations of the parties is not specified in the contract or if the contract states that the parties will agree to the price in the future, then such a contract is unenforceable by the courts.

Of particular interest to you, as a contractor, is that the failure of a contract to express specifically the terms of payment does not prevent that contract being deemed enforceable.

Contracts are enforceable that contain not only a fixed base price, but also a so-called escalator clause. An escalator clause states that in the event of a specified cost increase, the seller is allowed to increase his or her price by a predetermined percentage of the fixed base price.

In *Bethlehem Steel Co. v. Turner Construction Co.*, 2 N.Y.2d 456, 161 N.Y.S.2d 90 (1957), just such an escalation clause was held enforceable by a New York court. Bethlehem Steel was to furnish, deliver, and erect structural steel for a 20-story office tower being constructed by Turner Construction. The contract contained an escalation clause, providing that where "prices for component materials" were to increase or decrease, there would be a corresponding adjustment in the contract price. The adjustment was, in any case, limited to $15 per ton of steel. The court found that where an escalation clause is clear, unequivocal, and understandable when read in connection with the entire contract, then the clause is enforceable and no trial is necessary to determine the legal effect of the contract.

§ 1.2.3 Time

The law requires that the time to complete a contract be definite. If, however, there is no agreed-upon time for completion, the law will require that the contract be completed in a "reasonable" time. However, a contract requiring performance at a time to be agreed upon in the future is too indefinite to be enforced. However, a court considering all circumstances may be able to conclude that performance was to be completed in a reasonable time.

§ 1.2.4 Quantity

As a contractor, you will not so much be concerned with the quantity requirement in your construction contracts as in your contracts with material suppliers. The courts require that the quantity involved in an agreement be specified with some certainty. The courts have used the phrase "reasonable certainty." What criteria are used to determine whether a quantity term is reasonably certain? Where a quantity amount is not specifically stated, the method for ascertaining the amount should be provided in the contract itself. In the alternative, the courts will deem a contract quantity sufficiently definite if that quantity can be determined by interpretation based on admissible evidence.

1.3 OFFER

§ 1.3.1 Intent

If a contract is to exist, its existence is based on the intent of the contracting parties.

The courts will presume that a commercial transaction in a commercial setting, between business entities, under standard terms, demonstrates that the contract entered into is not unconscionable, and therefore avoidable. *Chrysler Credit Corp. v. Kosal*, 132 A.D.2d 686, 518 N.Y.S.2d 162 (2nd Dept. 1987). A binding contract exists not upon the subjective intent of either party, but upon the objective manifestations of intent. Those manifestations can be evidenced by actions as well as the words of the contract. *Wright v. Ford Motor Co.*, 111 A.D.2d 810, 490 N.Y.S.2d 556 (2nd Dept. 1985).

P. J. Carlin Const. Co. v. Whiffen Elec. Co., Inc., 66 A.D.2d 684, 411 N.Y.S.2d 27 (1st Dept. 1978) involved a dispute between a general contractor, who alleged that he had substantially reduced his bid for the construction of a high school in reliance upon an electrical subcontractor's representation by phone that he would perform electrical work for a certain price. After the general contractor received award of the project from the Greenwich Board of Education, the parties met face to face and agreed on a price, a scope of work. and security for performance. The subcontractor, however, maintained that the terms of the writings sent to the contractor were not discussed prior to the date of the writing, and that is was their understanding that there would be no agreement until a contract was signed by them.

The court acknowledged that the manner in which the parties negotiated the alleged subcontract was in conformity with the custom and practice in the construction trades. The court stated that a general contractor must often rely upon oral offers to perform subcontract work. The court felt that the determination of the existence of an oral contract, for the performance by the electrical subcontractor of work at the high school, must rest upon evidence at a full trial as to the intent of the parties, their expressed words and deeds, the totality of the acts of the parties, their relationship, and their objectives.

§ 1.3.2 Manifestation

The overt acts of a party must be considered in determining whether a contract exists. Conduct of a party is considered in determining whether it may be inferred that the party considers himself or herself under contract.

Recall the discussion at § 1.1.3. The acceptance of a contract, even with the ignorance of its contents, gives rise to a presumption of intent.

A court, however, will not bind a party to clauses printed on the reverse of a contract unless the evidence shows that the party was aware of those clauses and agreed to them. *Tri-City Renta-Car & Leasing Corp. v. Vaillancourt*, 33 A.D.2d 613, 304 N.Y.S.2d 682 (3rd Dept. 1969).

§ 1.3.3 Negotiations

Before a court will enforce a contract, it must appear that the contracting party against whom enforcement is sought agreed to the obligation in question. Parties may withdraw from contract negotiations with impunity. *Morse v. Swank*, 459 F. Supp. 660 (D.C. N.Y. 1978).

A contract cannot be established as long as the parties consider that an essential element remains open. The law will not make a contract where the contracting parties did not intend one. Where extrinsic evidence suggests conduct of negotiations rather than the establishment of a contract, the courts will not hold either of the parties bound.

Of course, as described in § 1.2, if the parties have provided a means for the court to determine their agreement or if the basic terms of the contract can be ascertained by resort to a reasonable standard, then the courts will uphold a contract.

§ 1.3.4 Agreement to Agree

Is an agreement to make a contract too indefinite to be enforced? The courts have often concluded that such is the case. Where an agreement to make a contract is incomplete in its essential charms, the courts have found such an agreement to be unenforceable. In *Briefstein v. P.J. Rotondo Constr. Co.*, 8 A.D.2d 349, 187 N.Y.S.2d 866 (1st Dept. 1959), the court found that Mr. Briefstein's allegations merely amounted to an agreement to make a contract. Since such an agreement is incomplete in its essential terms, it is unenforceable.

However, as discussed in § 1.3.1, an agreement to agree will not necessarily be unenforceable. The agreement may contain a provision that is left for future agreement.

May Metropolitan Corp. v. May Oil Burner Corp., 290 N.Y. 260 (1943), concerned a distributorship contract for oil burners. Although the number of units to be supplied from year to year was to be determined upon the annual renewal of the contract, the court found that this was more than an agreement to agree. The custom and usage in the trade, the course of conduct between the contracting parties over a number of years, and the question of the reasonableness of the proposed increase were all issues on which evidence should be heard. The distributorship's portion of the contract was, therefore, upheld.

In *Central Federal Savings, F.S.B. v. National Westminster Bank, U.S.A.*, 574 N.Y.S.2d 18 (1st Dept. 1991), the seller of a branch office of a bank sought to enforce a contract. The contract had allegedly arisen from an exchange of letters between the parties. However, the purchase was to be subject to "agreement on all terms and conditions relating to the proposed acquisition, to our satisfaction with all legal matters, and to the approval of our Board of Directors." The court noted that a draft agreement, sent to defendant National, by Central Federal, was over 30 pages long. In addition, there was no agreement on the essential term concerning payment for National's assumption of the Central Federal deposit liabilities. The court found that this issue had direct bearing on the price to be paid by National for the Central Federal branch. This could not be viewed as a situation where parties only had to formalize an agreement already reached. The exchange of letters and the content of those letters did not indicate the present intention to be bound, but rather an intention to negotiate essential terms of a contract. Since there was no "meeting of the minds," the court held that there was no evidence to support the existence of a contract. However, had the agreement indicated that the essential terms could have been determined by reference to extrinsic evidence, or that the parties considered themselves bound by the contract, then the court would have found it to be an enforceable contract.

§ 1.3.5 Invitation to Negotiate

In the construction industry, the concept of invitation to negotiate is most often encountered in the form of a bid proposal. An owner or public agency will make available plans, specifications, and its form of contract. As a contractor, you will be expected to take off the quantities, prepare your bid amount, complete the bid form, and present it at the appropriate time for bid opening. Most often, public agencies will require a bid bond. The bond guarantees that in the event the contractor does not accept the contract at the price he or she proposes, a surety company will make the owner whole for any loss he or she sustains. The bid proposal is interpreted by the courts as an invitation to enter into negotiations and not an offer that, if accepted by the contractor, would form a binding contract. The proposal merely solicits offers from various contractors. It is, in essence, a request for an offer.

Guarantee Const. Co. v. Rickert-Finlay Realty Corp., 88 Misc. 73, 150 N.Y.S. 551 (Queens Co. 1914) involved a dispute arising out of grading work performed in Flushing, New York. Guarantee Construction had written to Rickert-Finlay outlining a discussion held between a Guarantee Construction representative and the engineer for Rickert-Finlay. The letter concluded "we trust that the arrangement will be satisfactory, and upon your acceptance, will prepare a contract along these lines." Although the construction company claimed before the court that this was a contract, the court found that a plain reading of the letter made it perfectly clear that there was never a contract. The letter was a proposed agreement and suggestion. As such, it solicited an offer from Rickert-Finlay. The contractor's letter was a request for an offer, and not an offer that upon acceptance by Rickert-Finlay would result in a binding contract.

However, once an owner accepts the bid proposal from the contractor, that acceptance forms a binding contract.

The courts have repeatedly held that the solicitation of bids either through advertising, newspapers, or other media is merely a request for offers. A bid is nothing more than an offer. No legal rights are created until the offer has been accepted. Once those offers have been tendered by the contractors, the owner then has the option either to accept or reject any of them. More important, before an offer can be accepted, its terms must be definite and certain. *S.S.I. Investors, Ltd. v. Korea Tungsten Min. Co., Ltd.* 80 A.D.2d 155, 438 N.Y.S.2d 96 (1st Dept. 1981), aff'd 55 N.Y.2d 934, 449 N.Y.S.2d 173 (1982).

§ 1.3.6 Contract in Writing

An owner and a contractor may agree to all the terms of the contract, but may make the existence of the contract dependent upon its execution in writing. Generally, this can be understood as an extension of the theory of intent discussed at § 1.3.1. If the parties intend that the contract will not be given effect until it is executed in writing, or in a more formal instrument, that contract will not be given effect by a court. However, in spite of the parties' agreement to execute a more formal contract, if the parties act on the terms of the contract and conduct themselves as if they are bound by a contract, a court will find that a binding contract has arisen between the parties. As a general case, the mere failure to reduce an existing contract to writing will be insufficient to prevent enforcement of a contract.

However, you must exercise caution when dealing with state agencies. In *Belmar Contracting Co. v. State*, 233 N.Y. 189 (1922), the highest court of the State of New York found that until the state signs a contract submitted to it, the state cannot be liable for any damages suffered by the other contracting party. Belmar was the successful bidder for the improvement of a portion of a state highway. On August 25, Belmar was notified that it was the successful bidder, and directed to execute the contract. Belmar executed the contract and filed a bond on September 1. Before the contract could be executed, however, it needed the approval of the State Comptroller. The Commissioner of Transportation finally signed the contract on October 20. Because of the delay in signing, it was not possible to complete the work in the fall. The work was finished during the following year, and Belmar sought its increased costs of labor and materials. The court rejected a number of theories proposed by Belmar. The first was that the delay in formal execution of the contract was a breach of the underlying agreement for which damages may be recovered. The second theory was that the state had contracted to make the contract within a reasonable time, and had breached that agreement. Finally, there was the proposition that Belmar became bound under penalty of forfeiting its performance bonds, thereby demonstrating that these were not preliminary negotiations from which Belmar could withdraw. The court could not sustain an award under any of those theories. Precisely because Belmar was dealing with a public agency, the prevailing regulations and, in fact, the applicable highway law prevented liability from arising until formal execution of the contract. The court notes that had the state refused to execute the contract entirely, it would have incurred no liability. It also noted that such is equally true if it merely delays execution. *South Boston Iron Co. v. U.S.*, 118 U.S. 37. The court found no basis to sustain an award to Belmar.

Carroll-Ratner Corp. v. City Manager of New Rochelle, 54 Misc.2d 625, 283 N.Y.S.2d 218 (West. Co. 1967), aff'd 36 A.D.2d 795, 320 N.Y.S.2d 715 (2nd Dept. 1971), involved an electrical contractor suing the City of New Rochelle to have the Court award him a public improvement contract. Carroll-Ratner was suing the city for damages due to loss of profits. The court declined to allow the suit for lost profits to proceed.

A rejected bidder, in connection with a public improvement, cannot obtain damages because the statutes requiring the award of public contracts to the lowest responsible bidder were not enacted for the benefit of the contractor, but for the protection of the municipal corporation and its taxpayers. The statutes do not give a rejected bidder an action for their violations. An unaccepted bid cannot be the basis for a contractual relationship with a public agency.

While the litigation was pending, the City of New Rochelle awarded the contract to another. The issue most disturbing to the court was that the City, by awarding the contract to another, had rendered the court impotent to grant the relief sought by Carroll-Ratner.

Carroll-Ratner had argued that the City, by awarding the contract to another, breached a common law contract, which came into existence when Carroll-Ratner submitted its low bid. Carroll-Ratner argued that the City breached not the public improvement electrical contract, but a prior contract to award the electrical contract upon submission of the lowest responsible bid. The court rejected this argument and noted that no contractual relationship between a contractor and a public agency is created upon the contractor's submission of the low bid. Again, the court cautioned that an invitation to submit a bid is not an offer. The bid submitted would be an offer which, if accepted by a public agency, would form a binding contract.

Although the court was troubled by the city's award of the contract to another bidder while the action was proceeding, it could find no grounds to allow an award of damages.

The result would be different in Ohio. There, statutes do allow for recovery of damages if judgment be given for the contractor. *State ex. rel. Tucker v. Newark*, 10 Ohio Cir. Dec. 440 (Circuit Court of Ohio). The lesson, quite simply, is that where damages are based upon legislative enactment, the law governing damages will vary from state to state. What *Carroll-Ratner* demonstrates is that in the case of unscrupulous cities, it is better to be a contractor in Ohio than in New York.

In connection with the bidding of public contracts, the owner, a municipality or a state, is generally required to accept the bid of the lowest responsible bidders. Where, for whatever reason, the public agency declines to accept the lowest responsible bid, that bidder has recourse in the courts to force the public agency to accept its bid.

The right of an interested party to force a governmental agency to perform an action that it is required to perform stems from the ancient writ of mandamus. Pursuant to that writ, a public agency is mandated to perform required actions. The writ of mandamus is preserved in the United States Code at Title 28, § 1361. It is preserved in various state statutes and in New York State, in particular, it is known as an Article 78 proceeding, referring to the Article in the Civil Practice Law and Rules, which preserves the ancient writ. Of course, *Carroll-Ratner* teaches us that an injunction preventing the public agency from awarding the contract ought to be brought along with the action to compel the contract award.

In general, the courts will disregard an agreement to execute a more formal contract if it is the form alone, and not the substance of the agreement, which is subject to future approval. *Brassteel Mfg. Co. v. Mitsubishi International Corp.*, 21 Misc.2d 343, 192 N.Y. S.2d 200 (Kings Co. 1959).

In *Aces Mechanical Corp. v. Cohen Bros. Realty & Constr. Corp.*, 136 A.D.2d 503, 523 N.Y.S.2d 824 (1st Dept. 1988), modified without opinion, 531 N.Y.S.2d 218 (1st Dept. 1988), a mechanical contractor sought to enforce a contract by employing a back-dated letter of intent. The court found that there was no "meeting of the minds." The letter stated that the parties' agreement was subject to approval by the owner's engineers, with respect to the mechanical contractor's modification of the plans and specifications, and to "entering into a standard form of Agreement satisfactory to" one of the owner's partners and certain lending institutions. The letter of intent also stated that the price would depend on the final scope of work approved by the owner.

The court found no "meeting of the minds" and that the parties had only agreed to conclude a contract at some future date.

1.4 ACCEPTANCE

§ 1.4.1 Communication

The offer submitted by the contractor in the form of a bid proposal to the owner does not become a binding contract until the contractor receives notice of its acceptance. Notice of acceptance can range anywhere from a formal notice to proceed, received from a governmental agency, to a telephone call from the owner's engineer or architect. *Payne v. Lautz Bros. & Co.*, 166 N.Y.S. 844 (City Ct. Buffalo, 1916), aff'd 168 N.Y.S. 369 (Sup. Sp.T. 1918), aff'd 185 A.D. 904, 171 N.Y.S. 1094 (4th Dept. 1918).

§ 1.4.2 Time

If an offer specifies a time within which it must be accepted, that time frame must be adhered to. In the familiar public agency/contractor relationship, where the invitation for bid merely solicits offers from various contractors, it is the public agency that must accept the successful offer and thereby consummate the contract. This is the reason public agencies often request that contractors extend their bid for a specified period, often 30 or 60 days. Where the offer specifies no time within which to accept the offer, it must be accepted within a reasonable time.

Whether an acceptance was made within a reasonable time depends on all the surrounding facts and circumstances. *Berke v. Sherman*, 213 N.Y.S.2d 210 (Queens Co. 1961), aff'd 15 A.D.2d 548, 222 N.Y.S.2d 1019 (2nd Dept. 1961).

§ 1.4.3 Conduct

Naturally, an offer may be accepted by conduct alone (see § 1.3.2). In *Cortland Asbestos Products, Inc. v. J. & K. Plumbing & Heating Co.*, 33 A.D.2d 11, 304 N.Y.S.2d 694 (3rd Dept. 1969), a general contractor incorporated the bid of a subcontractor into his bid to an owner. The bid to the owner included both the amount of the bid from the subcontractor and the name of the subcontractor. The court held that that alone did not constitute acceptance. The contractor made no return promise to the subcontractor. Without a return promise, there is no acceptance of a construction contract and thus no contract. *Milone & Tucci v. Bona Fide Builders*, 49 Wash.2d 363, 301 P.2d 759; *Klose v. Sequoia Union High School Dist.*, 118 Cal. App.2d 636, 258 P.2d 515.

Acceptance cannot be inferred from a person's act and no contract arises where one of the contracting parties refuses the terms of the offer.

§ 1.4.4 Manner

If an offer specifies the manner in which it is to be accepted, acceptance is only effective if made in that manner. Acceptance cannot be ambiguous, qualified, or conditional. Acceptance by the usual means between two parties pursuant to their ordinary course of conduct will be effective as acceptance.

1.5 PARTIES

§ 1.5.1 Number

For a contract to be formed, there must be a "meeting of the minds." Implicit in that mandate is the general rule that a contract must have at least two parties. Without two parties, there can be no contract: A person cannot contract with himself.

§ 1.5.2 Capacity to Contract

The parties contracting must be recognized legally to form a contract. That is, one of the parties must have the ability to contract, and the other, the ability to enter into a contract. To establish valid agreement recognizable in the law, both parties must have the capacity to contract. *Buffalo Savings Bank v. Threeinwon Realty Corp.*, 175 Misc. 807, 26 N.Y.S.2d 324 (Sup. Ct. Erie Co. 1940).

1.6 CONDITIONAL ACCEPTANCE

§ 1.6.1 Differences

It is a basic premise of contract law, that the acceptance must comply with the terms of the offer. Should the terms of the acceptance of the offer differ, it is not considered an acceptance at all, but rather a counteroffer, which must then be accepted by the other contracting party. When an acceptance is qualified or when it places additional conditions on the offer, there can be no contract unless that qualification or the condition is accepted by the other party. *Warner Mfg. Co. v. Jacobs*, 177 N.Y.S. 183 (App.T. 1919).

Barber-Greene Co. v. M.F. Dollard, Jr., Inc., 239 A.D. 655, 269 N.Y.S. 211 (3rd Dept. 1934), aff'd 267 N.Y. 545 (1935), was an action to recover the purchase price of a trencher. The manufacturer's sales representative had received from Dollard, the contractor, a written order for the trencher. However, a contract, that was supposed to embody the terms of the agreement, was never signed by the contractor. The written order was subject to Barber-Greene's approval, and the order contained a provision that acceptance at Barber-Greene's home office was necessary to constitute a contract. Although Barber-Greene shipped the trencher to Dollard, Barber-Greene never formally accepted the order by signing it, nor was a notice of acceptance relayed to Dollard.

On October 11, the trencher arrived in Albany. Dollard notified the sales representative that the machine had arrived and requested the services of an expert mechanic. The mechanic did not arrive until October 18. Dollard then declined to receive the trencher, claiming an unreasonable delay.

Barber-Greene urged that although there was no written acceptance of the order, that upon shipping the machine, the order became a binding contract. Although acceptance may be indicated by conduct or acquiescence, acceptance must meet and correspond with the offer. The court found that the shipment alone did not constitute acceptance of the purchase order and offer. Without there having been acceptance of the Dollard offer, Dollard had the right to withdraw the offer at any time. Consequently, Dollard's notification that it refused to accept the machine constituted a withdrawal of its offer to purchase. No contract had arisen and there was no basis to find on behalf of Barber-Greene for breach of contract.

Differences with regard to the place of performance can prevent a contract from becoming effective. Where the offer states the place of performance, the acceptance must unequivocally and unconditionally accept that place of performance. However, when no particular place of performance is specified, the acceptance may suggest a place without invalidating the contract.

§ 1.6.2 Acceptance

Where an owner provides the contract, not all cards are stacked against the contractor. It is the contractor's right to accept the contract dependent upon any conditions he or she may wish. If the owner acquiesces to those conditions, the owner is bound by them. There will be no contract unless the owner agrees to the conditions set by the contractor.

Suppose an owner makes an offer that is accepted by the contractor subject to additional terms and conditions. The owner then proceeds as if a contract were in effect; i.e., by turning the site over to the contractor, or advancing payments for materials. A court will find that the contract formed incorporates the additional terms and conditions of the contractor.

American Tempering, Inc. v. Craft Architectural Metals Corp., 107 A.D.2d 565, 483 N.Y.S.2d 304 (1st Dept. 1985), involved a California manufacturer of construction glass. It was agreed that American would supply annealed and insulating glass and heat-strengthened laminated glass to Craft. American supplied all the materials and sought the outstanding balance. Craft's purchase order had stated that any action in connection with the purchase order must be brought within one year of the date material was last furnished. However, American's acknowledgment of the purchase order stated that if the terms of the acknowledgment differed from the conditions of the purchaser's order, then the acknowledgment would be construed as a counteroffer and not effective as an acceptance of the purchase order unless the purchaser, Craft, agreed to the terms of the acknowledgment. The acknowledgment would constitute the entire agreement between the parties. American's acknowledgment prevailed over the limitation in Craft's purchase order. American's acknowledgment was specifically conditioned on acceptance by Craft. Craft accepted by receiving the materials subject to the acknowledgment. Therefore, the action was not precluded by the limitation found in the Craft purchase order.

CITATIONS

Aces Mechanical Corp. v. Cohen Bros. Realty & Constr. Corp., 136 A.D.2d 503, 523 N.Y.S.2d 824 (1st Dept. 1988), modified without opinion, 531 N.Y.S.2d 218 (1st Dept. 1988) 1.3.6

Aluminum Fare, Inc. v. Abdella, 90 A.D.2d 603, 456 N.Y.S.2d 184 (3rd Dept. 1982), appeal denied 58 N.Y.2d 606, 460 N.Y.S.2d 1025 (1983) 1.1.2

American Tempering, Inc. v. Craft Architectural Metals Corp., 107 A.D.2d 565, 483 N.Y.S.2d 304 (1st Dept. 1985) 1.6.2

Application of Vincent J. Smith, Inc., 19 A.D.2d 763, 241 N.Y.S.2d 507 (3rd Dept. 1963) 1.1.2

Barber-Greene Co. v. M.F. Dollard, Jr., Inc., 239 A.D. 655, 269 N.Y.S. 211 (3rd Dept. 1934), aff'd 267 N.Y. 545 (1935) 1.6.1

Belmar Contracting Co. v. State, 233 N.Y. 189 (1922) 1.3.6

Berke v. Sherman, 213 N.Y.S.2d 210 (Queens Co. 1961), aff'd 15 A.D.2d 548, 222 N.Y.S.2d 1019 (2nd Dept. 1961) 1.4.2

Bethlehem Steel Co. v. Turner Construction Co., 2 N.Y.2d 456, 161 N.Y.S.2d 90 (1957) 1.2.2

Birch v. McNall, 19 A.D.2d 850, 244 N.Y.S.2d 60 (4th Dept. 1963) 1.1.1

Brassteel Mfg. Co. v. Mitsubishi International Corp., 21 Misc.2d 343, 192 N.Y. S.2d 200 (Kings Co. 1959) 1.3.6

Briefstein v. P.J. Rotondo Constr. Co., 8 A.D.2d 349, 187 N.Y.S.2d 866 (1st Dept. 1959) 1.3.4

CHAPTER 2
CONSTRUCTION

2.1 INTENTION

§ 2.1.1 Ambiguous Language

Where a court finds that a contract is unambiguous, the court need only look to the four corners of the document to determine the parties' agreement. Any evidence of acts, intentions, suppositions, and assumptions made by the parties other than those that exist in the contract itself will be ignored by the court.

The courts have held that where the meaning of contract language is plain, it cannot be interpreted as ambiguous merely because the parties urge different interpretations in litigation. The court cannot find a contract ambiguous merely because it is so urged by one of the parties where the interpretation would bend the contract language beyond its reasonable and ordinary meanings. Oral evidence of the meaning of a contract is only admissible where the contract language itself is ambiguous. *Burger King Corp. v. Horn & Hardart Co.*, 893 F.2d 525 (2nd Cir. 1990).

By what standard is ambiguity in a construction contract judged? The court would assume the place of a reasonably intelligent person who has examined the entire integrated agreement and who is aware of the customs, practices, usages, and terminologies in the construction trades. *Curry Road Ltd. v. K Mart Corp.*, 893 F.2d 509 (2nd Cir. 1990).

§ 2.1.2 Reasonable Interpretation

Where a contract is clear and unambiguous, a court can do no more than support a fair and reasonable interpretation of that contract. However, where the contract is less than clear, or some provision of the contract is ambiguous, the court must use a reasonable standard. The standard must uphold a meaning that would appear to reasonable persons at the time and place of the contracting.

Fleetash Realty Co. v. August Severio Constr. Co., 21 Misc.2d 350, 188 N.Y.S.2d 714 (West. Co. 1959), aff'd 11 A.D.2d 769, 205 N.Y.S.2d 212 (2nd Dept. 1960), involved an agreement between an apartment building owner and a contractor. The contractor asked for permission to remove any portion of the owner's fence in connection with construction of a garage building on adjoining property. The owner consented to any disturbance of the fence with the understanding that the contractor would indemnify and save harmless the owner from any claims arising from the work.

The owner conceded that the fence has been reset and restored to its original condition. In this lawsuit, the owner sought recovery of damages to the apartment building itself. Those damages were caused by blasting operations incidental to the contractor's construction of the garage building. The contractor argued that the indemnification was limited to the resetting and restoration of the fence.

The court found the contract between the parties clear and unambiguous. The contract referred unequivocally to the fence and not to the owner's apartment house. The court felt to hold otherwise would require straining the language of the contract. The suit against the contractor was dismissed.

If the intentions of the parties can be determined from the contractual language, then the courts must support an interpretation is conformance with that language. *Cromwell Towers Redevelopment Co. v. Yonkers*, 41 N.Y.2d 1, 390 N.Y.S.2d 822 (1976).

A contract that seriously disadvantages one of the parties will not be favored by the courts. The court would avoid such an interpretation in favor of a more reasonable one. A reasonable interpretation of a contract is considered one that gives parties relatively equal advantage rather than one that provides overwhelming advantage to one party over the other.

Tibbetts Contracting Corp. v. O & E Contracting Co., 15 N.Y.2d 324, 258 N.Y.S.2d 400 (1965), involved a dispute between an excavation contractor and a developer. While the ground at the site was frozen, the contractor moved machinery and equipment to another job. It was the intent of the contractor to return to the project when weather conditions and the progress of other work made its job easier.

The developer relied on various clauses that the court characterized as "draconian," including that the developer would decide all questions as to the performance, quantity, quality, acceptability, fitness, and rate of progress of the work, and that the work would be completed with all possible diligence and speed as otherwise directed by the developer. The court noted that all this appeared in a contract that had no completion date. The court found that the failure of the subcontractor to resume performance upon three days demand did not constitute a breach of contract. It considered that continuance or resumption of the work would have been futile or inordinately costly. The court found that the frozen condition of the work site would prevent backfilling and grading operations.

In another interpretation case, a general contractor was hired by the State to construct a dam and to excavate an emergency spillway. The contractor claimed payment for two items: one, constructing the dam; and two, for excavating the emergency spillway. The contract had provided that the materials for the dam embankment were to be obtained from a designated borrow area. The required site for the excavation of the emergency spillway had been designated as a "potential borrow source."

The contract provided that there would be no payment under the earth work item for material excavated from borrow sites or from the existing dam excavation. The contract further provided that the earth work categorically would not include excavation incidental to placing material in the embankment.

Other contract language could have been interpreted to categorize materials excavated from the emergency spillway area as previously excavated materials. This would allow payment to the contractor. The contractor's interpretation would necessitate the assumption that material excavated from the emergency spillway area was not borrow. The court held that the contractor's interpretation was unreasonable and found for the State.

The contractor's interpretation would necessitate two payments for what the court felt was essentially a single job. The material was to be moved from the emergency spillway area to the dam embankment. The parties had been aware at the time they entered into the contract that the dam embankment was to be constructed partially from material from the emergency spillway area. Thus, the emergency spillway area was designated as a borrow site on the plan. *A.J. Cerasaro, Inc. v. State*, 97 A.D.2d 598, 468 N.Y.S.2d 204 (3rd Dept. 1983).

In long and often complex construction contracts, the reasonableness of interpretation standard is often invoked by the courts. The import of seemingly conflicting clauses must be evaluated to compel a result that is both reasonable and which reflects the intent of the parties.

Finally, in *Deverho Constr. Co. v. State*, 94 Misc.2d 1053, 407 N.Y.S.2d 399 (Ct. Cl. 1978), a highway reconstruction contractor sought payment from the State for labor supplied to protect traffic and for watchmen patrols. Although the Comptroller had not approved the contract prior to certain of the labor being performed, that did not bar payments to the contractor. The court noted that since the State had prepared the contract, any ambiguity in the language must be resolved against the State. The interpretation would be what a contractor could reasonably understand the State's language to mean, not necessarily what the State had hoped it said. The court finally noted that the State's conduct in sending surveyors to the work site prior to the Comptroller's approval compelled the contractor to provide its labor for protection of traffic.

§ 2.1.3 Validity

Where the courts must consider an ambiguity in a contract, which could render the contract valid or invalid, the courts will generally favor an interpretation that finds the contract valid. The same general principle applies if the ambiguity could render the contract legal or illegal.

In general, the courts will find for the validity of the contract and for the legality of a contract, if such an interpretation is possible.

§ 2.1.4 Interpretation by the Parties

When the language of the contract is ambiguous, the courts will heed the conduct of the parties. The conduct of the parties are a sure indication of their intent. The parties' interpretation of the contract is given ample consideration by a court in determining the true meaning of the contract.

Beaver Engineering & Contracting Co. v. New York, 192 A.D. 662, 183 N.Y.S. 386 (1st Dept. 1920), aff'd 233 N.Y. 548 (1920), involved construction of a reservoir in Silver Lake, Staten Island. The contract provided for four classes of excavation: 1) for clearing and grubbing, 2) for removal of top soil, 3) for miscellaneous excavation, and 4) for shallow flowage excavation. The reservoir site was partially covered by a lake. The contractor claimed that material it was ordered to remove at the bottom of the lake should be paid for at the higher rate for miscellaneous excavation rather than the lower rate for shallow flowage excavation.

The court looked at the practical interpretation of the parties. In four prior monthly statements, this same excavation was classified as shallow flowage excavation by the City of New York. The contractor, Beaver, received the statements and made no objection that the classification was improper. The court found that the practical interpretation of the parties determined the question. It held that Beaver was only entitled to the unit price for shallow flowage excavation.

However, the courts cannot consider either the parties' interpretation or the parties conduct where the interpretation or the conduct clearly contradicts the plain meaning of the contract. The court will not construe a contract in such a way that it is inconsistent with the contract language itself. *Nau v. Vulcan Rail & Constr. Co.*, 286 N.Y. 188, reh'g. denied, 287 N.Y. 630 (1941).

§ 2.1.5 Surrounding Circumstances

When engaged in contract interpretation, a court will attempt to place itself in the shoes of the parties. It will look to the surrounding circumstances, to the parties' prior relationships, and to the actual language employed.

The courts will also be governed by their own perceptions of public policy issues. Where the court deems that an important issue of public policy is involved, the actual contract language or the parties' acts pursuant to the contract will not necessarily be given the weight they might receive in a situation where public policy considerations are absent. *Chard v. Ryan-Parker Const. Co.*, 182 A.D. 455, 169 N.Y.S. 622 (1st Dept. 1918).

Although the courts will consider extrinsic matters, in no event will the surrounding circumstances alone allow an interpretation of the contract that is unsupported by the contract language itself. As stated before, where the contract language is clear and unambiguous, evidence external to the four corners of the document will not be considered by a court. To clarify an ambiguity, however, the court is well within the limits of the reasonableness rule when it considers extrinsic matters such as the relationship of the parties and circumstances surrounding the negotiation and execution of the contract. All of those issues bear on the actual intent of the parties, whether the intent was identical, and whether a meeting of the minds did occur. *Broadway Maintenance Corp. v. New York*, 19 A.D.2d 96, 241 N.Y.S.2d 116 (1st Dept. 1963).

§ 2.1.6 Purpose of Contract

The determination of the parties' intent is the court's chief purpose in interpreting a contract. *Madawick Contracting Co. v. Travelers Ins. Co.*, 307 N.Y. 111 (1954). The intent of the parties as derived from the contract language is of primary concern to the court as it searches to determine the extent of the rights and obligations of the contracting parties. *Airco Alloys Division, Airco Inc. v. Niagara Mohawk Power Corp.*, 76 A.D.2d 68, 430 N.Y.S.2d 179 (4th Dept. 1980).

In the interpretation of the construction paragraph, not only the words of the contract but the nature of the work to be performed and the parties' conduct under the contract are considered.

The difficulty with construction contracts is the difficulty in the nature of construction. Changes in conditions, unanticipated work, and differing site conditions can all place a contractor and owner in situations not anticipated by the contract documents. In those situations, the court does not interpret the actual contract. If the intent of the parties can be determined from the contract language employed with respect to those situations that the parties did anticipate, then intent can be anticipated and deduced with respect to those situations that were not anticipated and have subsequently occurred.

Of course, as a general rule, the object of contractual interpretation by the courts is to do nothing more or less than give effect to the intent of the parties. The determination of the parties' intent is not governed by their unexpressed subjective views, but rather by their writing, their acts, their conduct, and all the circumstances surrounding the formation of the contract. Subjective intent is irrelevant. *Rock Ridge Townhouses, Inc. v. Village of Tupper Lake*, 99 A.D.2d 914, 472 N.Y.S.2d 521 (3rd Dept. 1984).

In the courts' view, the subject of a contract and the object to be advanced are to be heavily weighted in the determination of the parties' intent. *Tillman v. New York City Housing Authority*, 23 Misc.2d 598, 198 N.Y.S.2d 682 (N.Y. Co. 1960), appeal dismissed 13 A.D.2d 959, 218 N.Y.S.2d 936 (1st Dept. 1961), involved a suit for personal injuries brought against the New York City Housing Authority. The Housing Authority then sued its own contractor in connection with the work performed. The New York City Housing Authority contract contains expansive language of indemnification. The contractor is held to be the insurer of the Authority with respect to risks arising from acts and omissions of the contractor, of any subcontractor and of the Authority, whether the claims made by third persons occur before or after final payment. It is sufficient that the claims arise from the performance of the work.

However, this injury involved a stone slab, which the Authority had instructed the contractor not to remove. When construing this contract, in the light of the apparent object of the parties, the court felt that those parties could not have contemplated indemnification for injuries to third parties six months after the satisfactory completion of the job, especially where the Authority had instructed the contractor not to remove the stone, which caused the accident. Therefore, the complaint against the contractor was dismissed.

The courts will interpret the contract as a whole, determining the intent of the parties from four corners of the document rather than from sentences or clauses extracted and viewed in isolation. Every part of the contract must be interpreted to give effect to its general purposes. *Tougher Heating & Plumbing Co., Inc. v. State*, 73 A.D.2d 732, 423 N.Y.S.2d 289 (3rd Dept. 1979).

§ 2.1.7 Expressed Intent

When the contract language contains the expressed intent of the parties, that intent must be given full effect. However, the interpretation of contractual intent must be the interpretation of a reasonable man. Reasonable intent is not the subjective intent, which a party may propose. When the contract language is clear and unambiguous, the contract is interpreted by the court without consideration of surrounding circumstances or the subsequent acts of the parties.

To arrive at the reasonable meaning of the parties' words, the court considers the natural and ordinary meaning of the words of the contract.

The subjective meaning of the words are not considered. Should a party entertain a secret intent, such intent will not be considered by the court should it differ from the expressed intent.

A general contractor argued that he was entitled to a change order for providing glass in windows to be installed in a school building as part of a lump sum contract. The court dismissed the contractor's claim, stating that the explicit contract provisions provided for glass in the windows. The subjective intent of the contractor must fail where the explicit provisions of the contract provide otherwise. *Prote Contracting Co. v. Bd. of Ed.*, 132 A.D.2d 538, 517 N.Y.S.2d 425 (2nd Dept. 1987).

§ 2.1.8 Changing the Contract

A court cannot create a new contract by interpretation. A court also cannot change the written word to express better the intentions of the parties when that intention would contradict the written contract. *Ditmars-31' Street Development Corp. v. Punia*, 17 A.D.2d 357, 235 N.Y.S.2d 796 (2nd Dept. 1962).

Unless a contract is ambiguous, the court cannot, by exercise of judicial interpretation, add terms that it thinks the parties should have intended.

Often the courts will involve themselves in what is known as a "choice of law" question. Precisely which law applies to a given set of facts and a given contract? The question, in many cases, is answered by reference to contract documents. Public agencies and many private owners will insert a provision that provides that the contract is to be interpreted by the laws of a certain named state. Should such a provision be found in your construction contract, then the question of what law applies is simply answered. However, there are those contracts that contain no such provision. What law, then, does the court apply?

In determining which state's law to apply, the court will consider the place of the work to be performed, where negotiations were held, and the location of the contract signing. In a contract for grit blasting and coating, it was found that New York had the most significant relationship. New York law would apply since the negotiations and the performance were to occur in New York State. *Camrex Contractors (Marine) Ltd. v. Reliance Marine Applicators, Inc.*, 579 F.Supp. 1420 (D.C. N.Y. 1984).

2.2 INTERPRETATION

§ 2.2.1 Interpretation in Favor

The party who initiates the contract and brings it into being is primarily responsible for any inadequacy. The courts will interpret the contract to resolve any ambiguity in favor of the other party.

Steinhilber v. Challenger Steel Products Corp., 19 Misc.2d 770, 187 N.Y.S.2d 373 (Nassau Co. 1959), rev'd on other grounds 9 A.D.2d 695, 191 N.Y.S.2d 1009 (2nd Dept. 1959), involved a proposed exclusive dealership for a line of ladders. Pursuant to a letter addressed to Steinhilber from the manufacturing steel corporation, Steinhilber was to be the sole distributor for all of the United States and Canada. The steel corporation agreed to pay Steinhilber 10 percent selling commission on all orders procured by him. The orders were subject to being processed as to price, delivery, time, and establishment of proper credit by the buyers. The court found that the steel corporation that brought the agreement into existence was primarily responsible for its inadequacy. Therefore, the steel corporation should suffer for any shortcomings. The court's answer to the steel corporation's protest was that it was the steel corporation who proposed the contract in the first place. See also, *Dailey Co. v. Clark Can Co.*, 128 Mich. 591.

In general, the party who brings the contract into existence should justly suffer for its shortcomings.

§ 2.2.2 Interpretation Against

Where a contract contains an ambiguity or where contract provisions are in conflict, the courts will interpret the contract against the draftsman. *Evelyn Bldg. Corp., v. New York,* 257 N.Y. 501 (1931), involved a building affected by the construction of a subway under St. Nicholas Avenue. In the course of removal by the contractor of support along the front of the building, a landslide occurred due to a mud seam in the rock. The entire front of the building slipped into the subway excavation. The building owner sued the City of New York, who was constructing the subway, and the City sued its contractor. The City's contract contained various provisions dealing with indemnification of the City by the contractor and requiring the contractor to make good any damages. In interpreting the contract, the court held that since it was prepared by the City, any doubt or ambiguity would be construed against the City. The person supplying the form of the contract will have that contract construed against him or her.

In *Sklar Door Corp. v. Locoteta Homes, Inc.*, 33 Misc.2d 299, 224 N.Y.S.2d 294 (Nassau Co. 1961), the contract was found on plaintiff Sklar Door Corp.'s printed form. The payment provision on that form stated that Locoteta would pay Sklar the contract amount. Payment was to be made upon "completion of the installation and/ or delivery of the merchandise."

Locoteta built its first model home. It did not pay Sklar for the doors until some time after completion. Locoteta then built two more homes. Those doors were also paid for some time later. When Locoteta was about to build additional homes, Sklar declared it would not furnish any more doors unless it was assured it would be paid immediately upon delivery or installation. Upon review of the contract language, the court said that by reasonable interpretation, Sklar was to be paid upon completion of installation or delivery on all 25 homes affected by the contract. On the other hand, another reasonable interpretation would be that payment was to be made upon delivery or installation upon each home. Since ambiguity existed in the Sklar contract, the ambiguity was resolved against Sklar.

In *Diocese of Rochester v. R-Monde Contractors, Inc.*, 148 Misc.2d 926, 562 N.Y.S.2d 593 (Monroe Co. 1989), aff'd 561 N.Y.S.2d 659 (4th Dept. 1990), the Diocese of Rochester sought to recover for fire damage to a church. The damage was caused by the faulty installation of insulation around light fixtures. The architect sought to be relieved of liability by the court. The court focused on two provisions in the architect's contract. The first was that the contract did impose a duty to inspect the site, to be familiar with the progress and quality of the work. The court found that the architect's obligation to issue certificates of payment required the architect to be familiar with the quantity and the quality of the work performed.

The architect's contract did include a provision excluding the architect from responsibility for construction methods or for acts or omissions of the contractor. The court cited *Hunt v. Ellisor & Tanner, Inc.*, 739 S.W.2d 933 (Ct. Apps. Tx.), which held that identical language did not absolve an architect from liability. The owner, therefore, could hold the architect's responsible for these damages.

2.3 LANGUAGE

§ 2.3.1 Meaning of Language

The words of a contract are given their ordinary meaning unless the intent of the parties would indicate some different meaning.

Allied Chemical Corp. v. Alpha Portland Industries, Inc., 58 A.D.2d 975, 397 N.Y.S.2d 480 (4th Dept. 1977), dealt with the interpretation, in 1977, of a contract entered into in 1921. One of the contract provisions provided that the contract would terminate upon Alpha's ceasing to quarry stone from its Janesville quarry. In 1975, Alpha ceased quarrying stone on the land it had quarried since 1921. The court found that the contract had no applicability to lands that were not being quarried in 1921.

The meaning given to the words should be that which best conveys the parties' intent.

Words in a construction contract will be given a meaning based on the expectations of persons in the construction industry, established customs, and the course of dealings between the parties.

§ 2.3.2 Technical Words

Words used in the construction industry are given the meaning attributed to them by practitioners in the industry. Technical words are interpreted in their technical sense unless circumstances suggest a different interpretation.

Unambiguous technical words with a clear and definite meaning present no question of interpretation for a jury. In *Gotham Const. Corp. v. New York*, 225 A.D. 829, 232 N.Y.S. 563 (2nd Dept. 1929), the court found that the inscription on the contract drawings "mean high water elevation 100°° Transit Comm. datum" was not ambiguous. The notation had a clear and definite meaning understood by the engineers at the time they submitted the Gotham Construction bid. The court held that the notation signified the horizontal plane of mean high water at Sandy Hook as defined by the U.S. Coast and Geodetic Survey, which had been adopted by the Transit Commission as a datum from which vertical measurements were to be taken.

§ 2.3.3 Legal Meaning

Where parties use words that have an ordinary legal meaning, they will be bound by that legal meaning. For example, legally, a "condition" must occur before other consequences arise. It is not a condition as in the phrase "a differing site condition." However, if the parties have not intended their words to be given their customary legal meaning, they will not be given that legal meaning. The intentions of the parties are paramount.

§ 2.3.4 Meaning Understood

The meaning that the parties understand, the meaning that they intended, is the meaning that the courts will attribute to the contract.

In *Andrew Catapano Co. v. New York*, 116 Misc.2d 163, 455 N.Y.S.2d 144 (N.Y. Co. 1980), the dispute centered about contract provisions requiring Catapano to view and investigate the site. Catapano claimed that subsurface conditions, the severe slope of the rock formations, and other substantial deviations from the data provided by the City regarding excavation could not have been anticipated by review of the bid documents. In addition, these conditions could not be anticipated from inspection of the contract site. The City argued that Catapano should have anticipated these conditions based on a reasonable analysis of the data. The court felt that the conflicting factual assertions of the parties necessitated a full trial to determine whether or not the conditions found could have been anticipated from the information provided by the City.

2.4 IMPLICATION

§ 2.4.1 Good Faith

A condition of good faith and fair dealing is implied in every contract. The parties are obligated to cooperate in fulfilling the purpose of the contract. A party to the contract cannot seek to frustrate the other party to that contract in his or her performance of contractual obligations. The question of frustration of purpose is one that troubled the court in the *City of New Rochelle* case. See the discussion at § 1.3.6. However, in *New Rochelle*, the court ultimately found that there was no contract and, therefore, the principles of frustration of purpose did not apply. One result of the requirement of fair dealing is that an owner is bound to provide to a contractor the site of the work, although that is not specifically mentioned in the contract. In addition, an owner cannot, in any way, impede the contractor's work.

In *J.W. Brennan Const. Co. v. State*, 117 Misc. 816, 191 N.Y.S. 253 (Ct. Cl. 1921), the contractor was to improve a state and county highway in Chemung County. About 3.9 miles from the west end of the road, the highway crossed a ravine. A concrete culvert, to be built by the town, was to be installed at this point. Once the culvert was built, Brennan was to deposit earthfill, grade, and complete the improvement at that site. The town refused to proceed with constructing the concrete culvert. The State, although it was aware that the town was contesting the need for the concrete culvert, had nonetheless awarded the contract. The State assured Brennan repeatedly that the culvert would be completed. Brennan relied on the assurances of the State, as he had a right to do. The court awarded damages to Brennan as a result of the State's failure to provide the culvert site in a timely fashion.

The principle is explained in *Pernet v. Peabody Engineering Corp.*, 20 A.D.2d 781, 248 N.Y.S.2d 132 (1st Dept. 1964), which provides that the question of good faith is always one for interpretation by the jury rather than by the court alone.

In *Syracuse v. Dale Engineering Co.*, 233 A.D. 417, 253 N.Y.S. 444 (4th Dept. 1931), Dale was the successful bidder for a resurfacing contract with the City of Syracuse. The resurfacing material was to be asphalt applied over brick. The City Engineer directed Dale to apply an intermediary coat of Tarvia pursuant to an extra work order. Dale protested and attempted to prove to the City Engineer that because of a chemical reaction, the Tarvia would destroy the asphalt. The asphalt pavement went to pieces within a few months. The City sought to recover against both Dale and its surety.

Although the City Engineer had the legal right to order extra work of the contractor, the court felt that the case was clearly governed by the principle that no party to a contract will intentionally or purposely do anything to prevent the other party from carrying out its part of the agreement. See also *M.L. Ryder Bldg. Co. v. Albany*, 187 A.D. 868, 176 N.Y.S. 456 (3rd Dept. 1919).

Once the courts have determined that an agreement is a contract, they are bound to find that the contract imposes obligations of good faith, cooperation, and fair dealing.

§ 2.4.2 Law As Part of the Contract

Implicit in all contracts is the law in existence at the time and in the place of contracting. The obligations of the contractor will be determined by that law. *New York v. Interborough Rapid Transit Co.*, 257 N.Y. 20 (1931), stands for that proposition. Of special interest to construction contractors is the detailed history of the New York subway contained in this case. At issue was a proposal to increase the subway fare from the existing price of five cents.

The parties to the contract can, of course, modify the existing law, shorten statutes of limitation, or make the interpretation of the contract subject to a law of some other state. However, any such modification will, of course, be subject to the public policy of the state in which the litigation is brought.

§ 2.4.3 Omissions or Obvious Mistakes

In the existence of any agreement, there are certain implied promises, which are necessary to carry out that agreement. *Long Island R.R. Co. v. Northville Industries Corp.*, 41 N.Y.2d 455, 393 N.Y.S.2d 925 (1977), involved a license agreement by which Northville obtained the right to install and use pipeline along the Long Island Railroad's right of way. As a license agreement, the documents did not obligate Northville to construct and operate the pipeline. Such an interpretation would be contrary to the obvious intent of the parties.

While the license agreement did not obligate Northville to construct the pipeline, it did contain a provision providing for an annual minimum guarantee fee to the Railroad. This the Railroad was awarded.

The courts will insert missing words or correct the mistaken use of words in order to carry out the intent of the parties. The misdescription of an object will not allow one party to insist on a precise limited and technical interpretation of the term used. The court will look at the surrounding circumstances and the course of conduct of the parties.

§ 2.4.4 Implied Contract

A contract will be implied where an existing contract expires by its terms, yet the contracting parties continue to act as if the contract remained in existence. In the construction industry, this situation arises under two circumstances. In the first, a supplier of raw materials may continue to supply those materials even though the contract was to exist for a limited period of time. In the second, a subcontractor may be employed by a general contractor or by a construction manager to perform certain services over a limited time period and at a number of locations either unspecified or to be specified. Should the time duration of the contract expire by its own terms and the contracting parties continue to act as if the contract was in existence, the courts will so find.

2.5 CUSTOM

§ 2.5.1 Definition

The prevailing customs and usages of a trade or industry are incorporated into a contract unless the contract specifies otherwise. The custom should be well settled, reasonable, uniform, and not contrary to the rules of law or the intentions of the parties.

Customs are, of course, overruled by specific contractual provisions to the contrary. In *Valente v. Two Guys From Harrison, N.Y., Inc.*, 35 A.D.2d 862, 315 N.Y.S.2d 220 (3rd Dept. 1970), the issue was the percentage of compaction in a landscaping and grading contract. Where the unambiguous contract provisions called for a certain percentage of compaction, the customary trade usage is properly ignored by the court.

§ 2.5.2 Reasonableness

A custom or usage that the court deems unreasonable will not be enforced, unless the parties specifically agreed to be bound by it. *P.J. Kennedy & Sons v. Perkins & Squier Co.*, 154 N.Y.S. 101 (1st Dept. App.T. 1915).

However, there is a judicial presumption that any custom established by free and reasonable men is a reasonable custom.

§ 2.5.3 Trades

The custom or usage applied in construing a contract must be the custom or usage of the particular trade or industry under consideration.

The applicable customs of the construction industry are found in a number of cases.

Lucisano v. Paratore, 195 Misc. 45, 88 N.Y.S.2d 715 (Mun. Ct. Syracuse 1949), aff'd 198 Misc. 193, 98 N.Y.S.2d 608 (Onondaga Co. Ct. 1950), involved the construction of a brick house. The court found against the purchaser of the house. The purchaser claimed that a brick house must have exterior walls of solid brick, that is, two brick walls each four inches thick with a total thickness of eight inches. The building contractor claimed that the recognized usage in the local building trade established that brick house is built of one wall of brick over wood or hollow tile, but not of double brick. The court found that the contractor did provide a brick house when he provided a building of wood frame construction with an exterior brick wall four inches thick, laid over side walls of wood.

In *Novak & Rehner, Inc. v. State*, 192 Misc. 66, 78 N.Y.S.2d 577 (Ct. Cl. 1948), the question was whether a contractor, who had been engaged by the State to furnish labor and materials for heating work and service connections, was required by the contract to remove old equipment from the utility tunnel in which it was located. In no portion of the contract was there a clear statement that the contractor was required to remove from the utility tunnels the old piping and appurtenances. Here there was no testimony as to the usual custom or usage. Since the State drew the contract and did not direct that the existing piping and its appurtenances be removed from the utility tunnel, the court found for the contractor.

A person not engaged in a particular industry is not bound by the customs of that industry.

§ 2.5.4 Legality

Naturally, the custom of any industry cannot be contrary to law or to public policy.

§ 2.5.5 Antiquity

A custom does not develop overnight. To be valid, a custom must be generally known to those in a particular industry or trade.

The party against whom a custom is sought to be enforced must be aware of the custom or ought to have been aware of it with the exercise of reasonable and prudent care.

Cassin v. Stillman-Delehanty-Ferris Co., 185 A.D. 63, 172 N.Y.S. 754 (2nd Dept. 1918), involved a dispute between a contractor and a subcontractor. The issues involved who was obligated to furnish ladders for the use of construction employees. The subcontractor sought to prove a custom that general contractors furnish to roofing subcontractors ladders as a means of access to the roof. The court found that the evidence went no further than to show that roofing subcontractors often used ladders that a general contractor may have brought to the premises.

In *Heating Maintenance Corp. v. State*, 206 Misc. 605, 134 N.Y.S. 71 (Ct. Cl. 1954), the State Department of Education contended that a contract provision requiring the insulation of all duct work also required a contractor to insulate the plenum. Evidence introduced at trial indicated that the trade uses and custom show the plenum not to be a part of the duct system. The court concluded that the insulation of the plenum was not included in the contract and, therefore, awarded damages to the contractor.

In *George Colon Contracting Corp. v. Morrison*, 162 N.Y.S.2d 841 (Jefferson Co. 1954), aff'd 2 A.D.2d 869, 157 N.Y.S.2d 927 (4th Dept. 1956), appeal denied 3 A.D.2d 690, 158 N.Y.S.2d 797 (4th Dept. 1957), Colon attempted to prove that it was customary to allow a construction contractor to decide whether or not to suspend concrete pours during adverse winter weather conditions. The contract set forth a specific duration of 300 days. The alleged custom was in direct contravention of the expressed contract terms. The court held against the contractor. It might be noted that it did not help Colon that his chief witnesses indicated that should a contractor decide to suspend pouring, it was at the contractor's own risk and that the contractor's obligation to complete within the contract period was paramount.

2.6 CONDITIONS

§ 2.6.1 Generally

I find that the most difficult task in preparing an explanation of one discipline, the law, to another, the construction industry, is the transposition of various terms of art. When your lawyer speaks to you about conditions in the contract, he or she is not referring to, for example, differing site conditions, a concept well understood in both disciplines. The conditions your lawyer is discussing are those that must be in place prior to your assertion of subsequent rights under the contract.

The most dangerous contractual provisions are often found in the contracts of public entities. The contracts of the City of New York and its agencies are an example. Among those conditions about which you must be especially cautious are those that require that prior to filing any action against the City, a notice of claim must be filed. Because of the importance of this notice of claim requirement, it is discussed separately at §§ 2.14.7 and 2.15.6. Suffice it to say that the notice of claim provisions are not only often found in the contract itself, but also in the various public laws affecting both the municipality and its municipal agencies. Because you are dealing with a public entity, the courts construe the notice of claim provisions strictly. In spite of the onerous nature of the provisions, and I have heard jurors criticizing city attorneys for them, unless a notice of claim is timely filing, the contractor's claim, no matter how just and no matter what the sum, effectively disappears.

Another conditional provision against which you should be on guard is one that makes payment of funds down the contractual chain an express condition precedent for payment to you. In the most common case, a construction manager's contract will state that payment to you is contingent upon the construction manager receiving payment from the owner. When a conditional payment clause appears in a contract, you should immediately contact your attorney. Depending on the words used, the provisions can mean that you are to be paid within a stated number of days, within a reasonable time or, should the owner not pay the person with whom you have a contract, not at all. Because of the importance of these clauses, they are discussed at length at § 2.15.2.

As a construction contractor, you will most frequently deal with express conditions, those that are actually written out in your contract. There are, on the other hand, conditions that are called "implied." They are called implied conditions because the law implies them.

§ 2.6.2 Conditions Precedent

The best way to distinguish between ordinary conditions and conditions precedent is to say that the law requires that conditions precedent be performed before any subsequent obligations arise. The language that the construction professional must be aware of in reviewing his or her contract would be: "as a condition precedent ...," "it is an expressed condition precedent ..., " or "the condition precedent to this act is ... ."

The courts will interpret conditions precedents strictly. Should the conditions precedent not be performed, a court will hold that the obligations that flow from them simply do not exist.

Just what situations can be deemed conditions precedent?

Allis-Chalmers Mfg. Co. v. Malan Constr. Corp., 30 N.Y.2d 225, 331 N.Y.S.2d 636 (1972), involved a contract between Malan and the City of New York for the construction of a water pollution control project in Brooklyn. Allis-Chalmers was to provide Malan with air blowers, main sewage pumps, pumping units, accessories, spare parts and other appurtenant equipment. The relevant payment clause of the contract stated that payment by Malan to Allis-Chalmers was to be made only after payment by the City to Malan of installments pursuant to the City/Malan contract and in the ratio that the value of the Allis-Chalmers equipment delivered to Malan bore to the total value of equipment to be supplied by Allis-Chalmers or to the extent that Malan received payment from the City for Allis-Chalmers' materials, whichever was the lesser. Naturally, a dispute arose in connection with the payment clause. What particularly interests us here are the items that the court considers conditions precedent. Allis-Chalmers claimed that Malan had performed all conditions of the City contract and was therefore entitled to payments under that contract. Allis-Chalmers also claimed that it had delivered the materials required or specified in its contract with Malan. Allis-Chalmers claimed that all the materials furnished had been installed in the City project. It also claimed that Malan had received payments from the City, which included the payments for Allis-Chalmers' material. Finally, Allis-Chalmers claimed that Malan had received payment in full for all materials furnished, thus entitling Allis-Chalmers to the balance on its contract. This represents a lengthy list of conditions precedent. The very length and complexity of the list leads one to question whether these are all actually conditions precedent or merely contingent conditions.

§ 2.6.3 Concurrent Conditions

At law, concurrent conditions may be the most burdensome to perform, but not the most onerous. This is because concurrent conditions are interpreted as mutual conditions precedent. In the case of a concurrent condition, both parties enter into the agreement with their eyes open. There is no liability for either party until the condition occurs or is performed.

2.7 TIME OF PERFORMANCE

§ 2.7.1 Implication of Reasonable Time

Where a construction contract is silent as to the date of completion, a reasonable time for completion will be implied. Likewise, where there is an extension of time in that contract for an unspecified duration, a reasonable duration of time will be implied.

Senerchia Realty Corp. v. Yonkers Community Development Agency, 80 A.D.2d 889, 437 N.Y.S.2d 124 (2nd Dept. 1981), involved a litigation for breach of a deferred payment agreement. For reasons unmentioned by the court, the deferred payment agreement fixed no time for performance. The court held that where the agreement fixed no time, a reasonable time for performance must be inferred. The realty corporation had sought to have the lawsuit against it dismissed since the agreement contained no time limit for payment. The court found that the absence of a time limit is no basis for dismissal.

How long is an unreasonable period of time?

In *Brede v. Rosedale Terrace Co.*, 158 A.D.2d 494, 143 N.Y.S. 583 (2nd Dept. 1913), rev'd on other grounds 216 N.Y. 246 (1915), Rosedale Terrace Co. was a developer of suburban property. Brede had purchased certain lots from Rosedale. His contract provided that Rosedale would grade streets, plant shade trees, and put down cement sidewalks. The contract had no date of completion. The court stated that a reasonable time was implied and continued "six years is beyond the bounds of reason." A developer is required to live up to his or her expressed promise. Brede, continued the court, did not pay substantial sums of money for uncultivated farms and undisturbed sand lots.

But six months may very well lay at one end of the reasonableness spectrum. *Tupper v. Wade Lupe Constr. Co.*, 39 Misc.2d 1053, 242 N.Y.S.2d 546 (Schenectady Co. 1963), involved a contract for completion of a home by March 1, 1960. The contract contained no provision that time was of the essence. By June 1, 1960, the contractor stated that he would require three months to complete the construction. By October 1960, the structure was still incomplete. The court found that the contractor had breached his contract when the home was not completed by October 1960. The court cited a case that held the expiration of a month exceeded a reasonable time. See also *Leo Harmonay, Inc. v. Binks Mfg. Co.*, 597 F.Supp. 1014 (S.D.N.Y. 1984), aff'd 762 F.2d 990 (2nd Cir. 1985).

§ 2.7.2 Reasonable Time

A court's determination of reasonableness, as might be expected, is based on the particular facts and circumstances of the case. We are not talking about stopwatch precision here. However, leniency should not be expected where one of the parties is waiting for his or her home to be completed. While stopwatch precision is not the criteria for reasonableness, when we turn from our watches to our calendars the courts will hold that the standard of reasonableness has been violated.

It is difficult to apply a yardstick to the meaning of words. How long is reasonably long? An appellant court has held that "immediately" does *not* mean, as a matter of law, within a reasonable time. Immediately does not mean the next morning. *Goldblatt v. Longacre Const. Co.*, 170 N.Y.S. 54 (1st Dept. App.T. 1918).

Certainly a "reasonable time" contemplates a longer period than "as soon as possible" or "directly." On the other hand, in *Wetter v. Kleinert*, 139 A.D. 220, 123 N.Y.S. 755 (2nd Dept. 1910), the phrase "at once" meant within a reasonable time under the circumstances.

What is considered in determining a reasonable time? Certainly, the court or a jury will look at the subject matter of the contract and the intent of the parties making the contract. The status of the parties and the circumstances of the negotiations and performance are also taken into account.

§ 2.7.3 Time of Payment

Naturally, the interpretation of a payment clause is of utmost importance. Where the contract is silent as to the time of payment, the courts will infer a reasonable time. Most construction contracts, however, include fairly standard payment clauses. The payment clauses of concern here are those that general contractors will insert in their subcontracts. These clauses, known as "pay-when-paid" clauses, provide that the general contractor pays his or her subcontractors within a stated number of days after he or she receives payment from the owner or that the contractor's obligation to pay his or her subcontractors is limited to the funds actually received from the owner in payment for work performed by the subcontractors. As a matter of practical consideration, the courts have recognized that suppliers and subcontractors on a large construction project need prompt payment for their work and materials in order for them to remain solvent and in business. The general contractor will continue to be at risk for a delay in payment by the owner, where there is no express condition precedent in the contract, see § 2.15.2., transfering the credit risk for a delay in payment to the suppliers and subcontractors.

Across the country, in Massachusetts, New York, Maryland, North Carolina, and Oklahoma, the courts have construed pay-when-paid provisions to mean that the contractor shall pay his or her subcontractors within a stated number of days or a reasonable time thereafter. *A.J. Wolfe Co. v. Baltimore Contractors, Inc.*, 355 Mass. 361 (1969), *Schuler-Haas Elec. Co. v. Aetna Cas. & Sur Co.*, 49 A.D.2d 60, 371 N.Y.S.2d 207 (4th Dept. 1975), aff'd 40 N.Y.2d 883, 389 N.Y.S.2d 348 (1976), *Fishman Const. Co. v. Hansen*, 238 Md. 418 (1965), *Howard-Green Elec. Co. v. Chaney & James Const. Co.*, 12 N.C. App. 63 (1971), *Moore v. Continental Cas. Co.*, 366 F.Supp. 954 (W.D. Okl. 1973).

A curious variation of the payment problem occurred in *Georgia Glass & Metal, Inc. v. Arco Chemical Co.*, (Ga. App. Sept. 3, 1991). The owner of a construction project, at the insistence of a material supplier, had sent to the material supplier and the subcontractor a letter stating that the cost of the materials would be jointly paid to both parties. Once the owner filed for bankruptcy, the material supplier sued the subcontractor for payment. The subcontractor argued that the owner's letter constituted a pay-when-paid agreement. Had such an agreement existed, the subcontractor would not have been obligated to pay the materialman unless he was paid by the owner. The argument of the subcontractor was rejected by both the trial and appellate courts. The letter was merely a joint check agreement and did not affect the nature of the subcontractor's debt to his materialman. The subcontractor remained obligated to pay his materialman in spite of the owner's failure to pay.

§ 2.7.4 Duration

Where no duration is set forth in the contract documents, the courts will interpret that to mean that the contract should last a reasonable time. As a practical consideration, the courts will state that the contract may be terminated; i.e., its duration made certain, by notice to either party.

The disputes involving a contractor usually arise in connection with open-ended supply or maintenance contracts. Where the contract provides that the maintenance is to extend for an unspecified duration, the court will construe the duration to be for a reasonable period under the circumstances. *Metal Associates, Inc. v. East Side Metal Spinning & Stamping Corp.*, 165 F.2d 163 (2nd Cir., 1947).

2.8 TIME OF THE ESSENCE

§ 2.8.1 Express Provision

In general, if a contract is absolutely to be completed by a specified date, the contract must state that "time is of the essence," or other words to similar effect. If such a statement is not found, the courts will generally hold that, without cognizable prejudice resulting from delay, time is not of the essence. In inquiring whether the parties regarded time as an essential element of their bargain, the courts will look for a specific provision so stating. In a contract in which time is of the essence, that provision is strictly binding. A party cannot complain that such a provision is causing him or her hardship.

In a contract for a construction of a house, the parties, by modification, agreed to extend the time of performance for three months. The court determined that such an extension of time indicated that the parties considered time to be an essential factor of the agreement and, therefore, strictly held the contractor to the extended date. *Tupper v. Wade Lupe Constr. Co.*, 39 Misc.2d 1053, 242 N.Y.S.2d 546 (Schenectady Co. 1963).

When a contractor is faced with a "time is of the essence" clause, he or she should realize that the courts will grant no leeway should damages occur.

§ 2.8.2 Effect of Demand

When a contract does not specify that time is of the essence, after an elapse of a reasonable amount of time, a contracting party can subsequently make time of the essence by giving sufficient notice.

Notice making time of the essence is especially useful in a situation where the completion date has been waived by one of the parties. Before time can again be made of the essence, the party in default must be given reasonable notice that performance is to be completed within the reasonable time specified in the notice. The notice should also state that if performance is not completed within this time, the contract will be at an end. This procedure may be used whether or not the original contract stated that time was of the essence. *Henderson Tire & Rubber Co. v. Wilson & Son*, 235 N.Y. 489 (1923).

§ 2.8.3 Effect of Character

The courts consider the particular character or nature of the contract in determining whether time is of the essence. In commercial contracts and in option contracts, time is generally presumed to be of the essence.

Federal Terra Cotta Co. v. Potterton Bros., 172 A.D. 705, 159 N.Y.S. 121 (1st Dept. 1916), aff'd 225 N.Y. 623 (1918), involved the delivery of terra cotta for incorporation in a building facade. Federal agreed to manufacture certain terra cotta according to the plans and specifications and to deliver a portion of the terra cotta to Potterton Bros. within 56 days after April 16. Federal was to receive approved shop drawings and selection of colors by Potterton on April 16. The color selections were made and the shop drawings were received by Federal on April 18. In order to begin the manufacturing process, Federal also needed to know the type of surface to apply to the terra cotta. The surface type was not selected until April 25.

Federal argued that since it had not received all information by April 16, it was relieved of its obligation to make the first delivery within 56 days. The court felt that Federal's position could only be supported if the contract was construed to make receipt of all information by April 16 a condition precedent to delivery within 56 days. The court found no such obligation under the contract and found that the delay in providing all information until April 25 did not relieve Federal of its obligation to provide the terra cotta within 56 days. Federal should have delivered the terra cotta 56 days after April 25, once it had received all necessary information. The court found that time was of the essence in this contract where both parties knew that Potterton Bros. would be subject to a penalty if it failed to complete the building on time.

In *People v. Metropolitan Surety Co.*, 168 A.D. 615, 154 N.Y.S. 373 (3rd Dept. 1915), appeal dismissed 217 N.Y. 612 (1916), time was deemed not to be of the essence in connection with a contract for mason, stone, and cement work. The contract provided that the entire building should be ready for occupancy on or before a certain date. However, there was nothing in the contract specifying just when the mason, stone, and cement work was to be done or at what stages ancillary work was to begin. The contract merely stated that work was to be performed in such a manner as not to delay other prime contractors in performing their work by the occupancy date. The court felt that the contract was not drawn with the specificity required for time to be of the essence.

§ 2.8.4 Specification of Date

In building construction contracts, the general rule is that once the date of performance is fixed by the contract itself, then time is considered of the essence. Where a construction contract is to be performed by a specified day, time is of the essence. *Milkolajewski v. Pugell*, 62 Misc. 449, 114 N.Y.S. 1084 (Erie Co. 1909).

In *Q.R.S. Co. v. Phillips-Jones Corp.*, 194 A.D. 170, 185 N.Y.S. 127 (1st Dept. 1920), aff'd 233 N.Y. 626 (1922), the contract was between a manufacturer and a land owner. The manufacturer was to obtain permits and construct a switch track over the owner's land within six months. The court found that the six month time limit was established to fix a date when rent would commence. As such, the time limit was insufficient to establish that time was of the essence in this contract. The manufacturer could delay construction beyond the six months without forfeiture of his rights under the contract.

The court considered the circumstances of the parties. The manufacturer had worked assiduously and diligently in obtaining the necessary permits and has expended over $670,000 in the purchase of additional lands and the erection of buildings thereon. The monies were expended with the knowledge of the land owner and reliance on the contract with the land owner. The manufacturer had partly performed on its part of the agreement and had expended a large sum of money relying upon the land owner's obligation. The court found that it would be a fraud not to allow the manufacturer the benefits of the contract.

Rochester v. LeCesse Bros. Contracting, Inc., 101 A.D.2d 1022, 476 N.Y.S.2d 703 (4th Dept. 1984), aff'd 64 N.Y.2d 1082, 489 N.Y.S.2d 905 (1985), involved a dispute between LeCesse and the City of Rochester. There was an inconsistency in the contract documents concerning the completion date. A proposal stated one date. Yet the contractor had agreed to complete by a date specified in the Supplementary Instructions to Bidders. The general conditions of the contract state that the Supplementary Instructions should take precedent over all contract documents other than addenda. The court held that the date in the Supplementary Instructions were to govern.

2.9 PERSONS BOUND

§ 2.9.1 Persons Not Named

The general rule, both in law and in common sense, is that the rights and obligations of a contract apply only to those persons making the contract. Only the parties to a contract can breach that contract. The parties to a contract cannot impose any liability on another who is a stranger to the contract. *New Amsterdam Casualty Co. v. Parsons*, 219 A.D. 486, 220 N.Y.S. 340 (3rd Dept. 1927).

A person whose signature does not appear on the contract and who is not named in the body of the contract is not bound by that contract. However, even if a person is not named in the body of the contract, if he or she signs the contract, he or she will be liable on the contract.

An owner was brought into a suit involving a prime contractor and its subcontractor. The contract between the prime and its subcontractor indicated that it did not create any contractual liability between the subcontractor and owner. There was no evidence presented that the owner was in any way under a contractual liability to the subcontractor. Neither did the owner assume any obligations to pay the subcontractor by its actions. Since the owner was a stranger to the subcontractor, the subcontractor could not move directly against the owner for payment under the contract. *Delta Electric, Inc. v. Ingram & Greene, Inc.*, 123 A.D.2d 369, 506 N.Y.S.2d 594 (2nd Dept. 1986).

Blitman Constr. Corp. v. Kent Village Housing Co., 91 A.D.2d 173, 457 N.Y.S.2d 817 (1st Dept. 1983), involved a breach of contract action by Blitman against Kent Village, a limited-profit housing agency and against the New York City Department Housing Preservation and Development (HPD). HPD was the supervisory agency. Among its responsibilities were the management and control of limited-profit housing companies, such as Kent Village. The court determined that HPD was not a party to the contract, although HPD had executed loan and mortgage agreements with Kent Village.

Although Blitman alleged purported acts of ratification that would subject HPD to contract responsibilities, the court found that those acts were purely supervisory acts.

The court found that the supervisory acts of HPD could not be considered unauthorized acts of control or interference, which could make HPD accountable under the contract. The facts in this case showed that Kent Village was to bear the responsibilities for any breach of contract. Unlike municipal contracts, this contract was not publicly bid.

The court determined that this was a private contract. The parties to the contract, Blitman and Kent Village, could not involve HPD, a stranger to that contract, in their dispute.

2.10 COMPENSATION

§ 2.10.1 Payment

As you must now realize, the courts will not attempt to infer a reasonable price. The price or compensation for a contract must be contained in the contract. If it is not, the contract is unenforceable by the courts. This is not a problem that generally arises in construction contracts.

The purpose of this section is not to give a detailed explanation of every potential pitfall in payment provisions. Instead, I will provide a brief overview of the court's interpretations of various contractual payment provisions. Other sections of this volume contain a detailed analysis of important payment provisions. For time of payment, see § 2.7.3. For contingent payment provisions, see § 2.6. For conditions precedent for payment, see §§ 2.6.2 and 2.15.2. Payment for extra work is discussed at§ 2.13.

Although the courts will not infer a price, naturally, in construing an interpretation of price, the courts will employ a reasonable standard that comports with the intent of the parties. *Bethlehem Steel Co. v. Turner Constr. Co*, 2 N.Y.2d 456, 161 N.Y.S.2d 90 (1957), involved a subcontract by which Bethlehem Steel was to furnish, deliver, and erect structural steel for a 20-story office building. Turner Construction was the general contractor. The contract clause provided that should "prices for component materials" increase or decrease, there would be a corresponding increase or decrease in the contract amount. Bethlehem contended that "prices for component materials" included the price of steel being regularly charged to the trade. In this case, that price had been increased by ten dollars per ton for steel shapes, plates, bars, sheets, rivets, and bolts. Bethlehem's adversary contended that "prices for component materials" only had reference to changes in price for materials used in progressing the work, such as increases in the price paid by Bethlehem to obtain materials necessary to produce steel. This interpretation held that the changes in the prices for component materials would reflect that cost to Bethlehem in the manufacture of raw steel at the mill, based on the elements employed in the manufacture, such as iron ore, steel scrap, and limestone.

The lower court held that an issue of fact was presented. An appellate court reversed and found no issue of fact. The highest court of New York found that no trial was necessary, since the words were not ambiguous. The price adjustment clause made specific mention of the basis for computing changes in the steel price.

The court felt that if "component materials" were used to mean those raw materials employed in production of steel, then the escalation on account of labor rates would not be limited to those specifically mentioned in the contract for design fabrication and erection, but would have included labor costs at the mill. The contract did not. The clause specifically stated "prices for component materials, labor rates applicable to the fabrication and erection thereof and freight rates." The court found that this language was clear in and of itself.

A general contractor had delayed payment to a subcontractor since the owner had not yet paid the general contractor. The general contractor could not show that the owner's delay in payment was due to any purported acts or omissions on the part of the subcontractor. The court found that the subcontractor was, therefore, entitled to payment of its retained percentage. *Grossman Steel and Aluminum Corp. v. Samson Window Corp.*, 78 A.D.2d 871, 433 N.Y.S.2d 31 (2nd Dept. 1980), aff'd 54 N.Y.2d 653, 442 N.Y.S.2d 769 (1981).

Michael J. Torpey, Inc. v. Consolidated Edison Co. of New York, Inc., 99 A.D.2d 484, 470 N.Y.S.2d 426 (2nd Dept. 1984), appeal dismissed 66 N.Y.2d 915, 498 N.Y.S.2d 1027 (1985), reconsideration denied 67 N.Y.2d 702, 499 N.Y.S.2d 941 (1986), involved an excavation contractor's complaint that the work assigned grossly deviated from "typical layouts" included with the invitation for bids. However, the contract contained language stating that the estimates were given solely as estimates and were not to be considered as guarantees of amounts or even as approximate amounts, which Con Edison would purchase. The contract further stated that whenever estimated quantities for unit price appeared in the contract, they were given solely for use in comparing bids. Con Edison expressly reserved the right to increase or diminish the amounts. In addition, the contract contained language that an increase or diminution would not give cause for any damage claim.

The court found that the contractual language was clear and unambiguous. Torpey claimed that it would produce evidence of custom and usage between Con Edison and other contractors that actual work under excavation contracts was to resemble the depiction used in the "typical layouts." Because of the extensive exculpatory language in Con Edison's documents, the court dismissed Torpey's complaint. The court found that evidence of custom and usage cannot be used to contradict or alter expressed terms of an unambiguous contract. Therefore, the court found no triable issues of facts.

In *McGowan & Connolly Co. v. Kenny-Moran Co.*, 207 A.D. 617, 202 N.Y.S. 513 (1st Dept. 1924), the contractor attempted to recover compensation additional to that in the terms of the written agreement. The court held that the contractor would have the burden of proving that the original contract was terminated and that a subsequent contract was made, allowing for the increased compensation.

Uvalde Asphalt Paving Co. v. New York, 196 A.D. 740, 188 N.Y.S.304 (1st Dept. 1921), involved a contract for the construction of the Grand Concourse from 161st Street to Mosholu Parkway in the Bronx. Uvalde Asphalt had complied with improper demands of New York City under protest and had completed its contractual work. The court found that payment for the protest work must be based upon the reasonable value of the increased costs incurred. The court pointed out that Uvalde was not prevented from performing its contract, which would make it impossible for him to show what the actual cost of the work was. Where the contract is not performed, the measure of the contractor's damage is the difference between his or her estimated cost to perform and the contract amount; that is, his or her lost profit. In *Uvalde*, the protest work was performed and the cost of performance must be actual, rather than estimated.

Hayes v. American Bridge Co., 127 A.D. 576, 111 N.Y.S. 883 (1st Dept. 1908), involved a salesman's contract with a predecessor of American Bridge Co. The contract stated that Hayes was to receive one-half of the profits on contracts he obtained for structural bridge work. In construing that contract term, the court found that profits meant actual profits realized rather than estimated profits. In a contract where compensation is a percentage of profits, where no profits are realized, there is no compensation due.

A variation of a contingent payment clause arose in *Lowy and Donnath, Inc. v. New York*, 98 A.D.2d 42, 469 N.Y.S.2d 760 (1st Dept. 1983), aff'd 62 N.Y.2d 746, 476 N.Y.S.2d 830 (1984). The court found that an electrical subcontractor was entitled to receive payment for the labor and materials furnished to the general contractor. The purchase order form of subcontract between the general contractor and the electrical subcontractor made the subcontract subject to the contract between the general contractor and the owner, to the extent that the electrical work was to be in strict accordance with the contract plans and specifications. Absent from the language was any indication that payment for the electrical work was subject to verification or audit by the public agency or payment from the owner to the general contractor. In addition, the purchase order did not contain any provision regarding the timing of payment. Therefore, payment was due immediately upon completion of the work.

Often, an owner will insert a clause in a contract requiring the general contractor to completely pay all subcontractors as a condition precedent to final payment. Where the general contractor was suing the owner for progress payments, the existence of such a clause would not prevent the judgment for the general contractor. *John W. Cowper Co. v. Buffalo Hotel Dev. Venture*, 72 N.Y.2d 890, 532 N.Y.S.2d 742 (1988). Public contracts often contain a provision that designates a certain municipal official to resolve disputes as to whether work is included in the contract or is extra work. Naturally, the municipality will pay for extra work, but not for work that it determines is in the contract. In *Kalisch-Jarcho, Inc. v. New York*, 72 N.Y.2d 727, 536 N.Y.S.2d 419 (1988), the contractor argued that owing to collusion between municipal officials, the contractor could be required to perform work blatantly beyond the scope of the contract, yet would be unable to collect for the work. Although the court acknowledged the problem, it stated that the work in question in this particular case was not so manifestly beyond the scope of the contract. Therefore, the court concluded there was no threat to the public interest.

As a contractor, you should review payment provisions carefully and contact your attorney for an explanation of precisely how the courts have interpreted these provisions in the past.

For particularized services, such as surveying, limited engineering analysis, and small construction contracts, payment is often pursuant to a "cost-plus" provision. Certain defined costs are paid for, as well as an agreed percentage of those costs.

In *Westinghouse, Church, Kerr & Co. v. Long Island R. Co.*, 160 A.D. 200, 145 N.Y.S. 201 (1st Dept. 1914), aff'd 216 N.Y. 697 (1915), the contract called for Westinghouse to electrify the Long Island Railroad. The compensation for design work was five percent of the actual cost of the completed work. The compensation for all other services was the actual cost incurred, with ten percent added. The contract defined actual cost to include all disbursements for machinery, apparatus, materials, supplies, labor and service, and miscellaneous charges, such as fire and casualty insurance, cost of accidents and breakages.

The implication of good faith applies to "cost-plus" contracts, as well. Where work is improperly done, the contractor is not entitled to charge for correcting and performing that work properly. *Hoggson Bros. v. Spiekerman*, 175 A.D. 114, 161 N.Y.S. 930 (1st Dept. 1916).

Hamilton v. Coogan, 7 Misc. 677, 28 N.Y.S. 21 (N.Y. Co. 1894), aff'd 148 N.Y. 753 (1896), involved a cost-plus contract which stated that costs were to be paid for and ten percent added as profit. The court held that the additional ten percent was to be paid on the entire amount paid to subcontractors, even where the subcontractors did their work on a cost-plus basis.

Albert A. Volk Co. v. Cauldwell-Wingate Co., 272 A.D. 290, 70 N.Y.S.2d 662 (1st Dept. 1947), involved a contract for the demolition of existing structures and erection of a new building on the site. This cost-plus contract contained an important provision. The person, either owner or contractor, paying for the work must have some way of determining the actual cost incurred. The provision inserted here allowed the general contractor to audit the books of his subcontractor and authorized the general contractor to check directly with the subcontractor's suppliers of labor and materials in order to verify the charges for labor, materials, and other items appearing in the subcontractor's bills. A prudently drafted cost-plus contract will have such, or similar, verification procedures. As the person subject to such verification, you, as contractor or subcontractor, must recognize that a review and audit of your books will be forthcoming.

In *Atlantic, Gulf & Pacific Co. v. Woodmere Realty Co.*, 156 A.D. 351, 142 N.Y.S. 953 (2nd Dept. 1913), appeal dismissed 209 N.Y. 557 (1913), the court held that the owner in connection with the cost-plus construction of levees along a canal could not complain of the expense. The owner was in the best position to ascertain the soil characteristics along the canal. The high cost was due primarily to the soft yielding characteristics of the soil itself, which moved out under the weight of the fill placed behind the levees. Concomitant lateral pressure exerted by the fill itself was greater than anticipated due to a high proportion of liquid contents.

The court pointed out that these are ordinary engineering problems. Woodmere Realty, the owner, could have determined both the stability of the subsoil and the load the soil would carry. The owner cannot now attempt to place the burden of these difficulties on the contractor.

2.11 COMPLETE CONTRACT

§ 2.11.1 Entire Contract

The court's goal in construing a contract is to arrive as closely as possible at the original intent of the parties. Each provision of the contract must be given its full force and effect. Where there exists an apparent conflict between provisions in a contract, it is the duty of the court to reconcile those provisions if possible.

The court must examine the entire contract and consider the relationships of the parties and the circumstances under which the contract was executed. Particular words should not be considered in isolation but in light of the obligation as a whole. Form should not prevail over substance. A sensible meaning of words should be sought in light of the intentions of the parties. *William C. Atwater & Co. v. Panama R. Co.*, 246 N.Y. 519 (1926).

Contracts often contain an obligation by the subcontractor to indemnify the general contractor for any liability imposed on the general contractor for property damage due to an act or omission of the subcontractor. In *Madawick Contracting Co. v. Travelers Ins. Co.*, 307 N.Y. 111 (1954), the subcontract required a subcontractor to obtain insurance coverage for such contractual indemnification. The subcontractor told his agent what was required by the contract and presented a copy of the passage to him and read it to him. The insurer then issued a certificate, which substantially set forth the indemnity clause of the contract. When the general contractor made a claim against the subcontractor and served a demand for arbitration upon the subcontractor, the insurer refused to defend. The insurer's refusal was based on its claim that the certificate only required it to defend a "suit" against the subcontractor.

The court held that the insurer was obligated to defend the subcontractor in an arbitration proceeding. The court found that the object of the insurance policy was to indemnify the subcontractor against liability resulting from his operations. Travelers, having incorporated the relevant portions of the subcontract, must be deemed to have implemented that clause in its context as part of the entire subcontract. Therefore, since arbitration was the exclusive remedy under the subcontract, Travelers Insurance cannot escape liability, claiming it was ignorant of the arbitration provision. The contract of an insurance company, in connection with a building or construction contract, must be construed together with the other instruments to which it refers.

§ 2.11.2 Effect of All Parts

Every provision of a contract should be given effect. A court will seek that interpretation, which allows a meaning for each provision of the contract. *Muzak Corp. v. Hotel Taft Corp.*, 1 N.Y.2d 42, 150 N.Y.S.2d 171 (1956).

Where, on the face of a contract, there is no indication to the contrary, words should be taken in their ordinary and grammatical sense. *In Re Thompson's Estate*, 18 Cal. App.2d 680 (1937), holds that an appropriate interpretation of a contract considers the entire document, reads the various parts together, and gives effect to each one. Each provision is used to aid interpretation of the other provisions. *Security-First Nat. Bank of Los Angeles v. Reynolds*, 104 Cal. App.2d 697 (1951). The rights of parties to a contract must be determined using the plain meaning of the language. The courts have said that the intent of the parties must be gleaned from the four corners of the document, rather than from isolated clauses or sentences. Every part of the contract should be interpreted to give effect to its general purpose. Where a literal interpretation would defeat and contravene the purpose of a contract, such an interpretation should be avoided. *Tougher Heating & Plumbing Co. v. State*, 73 A.D.2d 732, 423 N.Y.S.2d 289 (3rd Dept. 1979). The courts seek to arrive at a fair and reasonable interpretation and will not allow a party to take advantage of "adroit and slippery language" in a contract. *Application of Seneca Falls Central School District*, 117 Misc.2d 879, 459 N.Y.S.2d 689 (Seneca Co. 1983).

A court will not rewrite an unambiguous contract. The court must enforce the contract's plain meaning. *C.C.G. Associates I v. Riverside Associates*, 157 A.D.2d 435, 556 N.Y.S.2d 859 (1st Dept. 1990).

§ 2.11.3 Conflicting Clauses

Sam Berley Heating Corp. v. Bd. of Ed., 206 N.Y.S.2d 38 (Mun. Ct., N.Y.C. 1960), involved a contract for the installation of new convectors, steam piping, and appertenant equipment at a public school. Berley contended that the contract only required it to cover with insulation and encase with sheet metal the convector connections in the corridors and stairways of the school, but not in the classrooms. The court performed an extensive analysis of the Schedule Specifications and the Standard Specifications, which formed a part of the contract. The court noted that, in construing contracts, all provisions must be reconciled and given effect, if such can be done. Any conflicts between the different provisions are not to be presumed if those provisions can stand together. The court felt that Berley's interpretation did violence to the provisions of the contract and attempted to create conflicts where none existed.

The court concluded that the contract provided that all convector connections in corridors and staircases shall be enclosed in #20 gage galvanized steel and attached to wall or baseboard. In addition, each connection shall be covered throughout its entire length with magnesia sectional covering and then encased in sheet metal casing. Where so specified, horizontal steam and return pipes near floors were to be encased in #20 gage galvanized steel fastened to floorboard. The contractor was required to encase in sheet metal all pipes connected to convectors, not only those in hallways and stairways, but those in classrooms, as well.

It is the court's duty to find harmony between conflicting provisions. However, where two clauses cannot both exist, the courts will allow the first provisions to stand and the subsequent provision to fail.

§ 2.11.4 Recital

Certain construction contracts may contain what is called a *recital*. The recital would indicate the background, motives, and purposes of the parties. The recital is generally not considered a part of the actual contract; however, it will aid the court in determining the parties' intent. If a recital conflicts with contractual provisions, the contractual provisions must prevail. However, where the recital is clear and an operative condition is ambiguous and inconsistent, the recital clause will govern.

§ 2.11.5 Written Material

Construction contracts often contain written or typewritten provisions. Is there a priority between those provisions and printed provisions of the agreement? The courts will hold that handwritten or typewritten provisions represent the expressed manifestations of the parties' intent. Those provisions will take precedent over any inconsistent provisions on the printed form. Of course, that is only the case where there is a conflict between the handwritten or typewritten provisions and the actual printed provisions. *Cale Development Co., Inc. v. Conciliation and Appeals Bd.*, 94 A.D.2d 229, 463 N.Y.S.2d 814 (1st Dept. 1983), aff'd 61 N.Y.2d 976, 475 N.Y.S.2d 278 (1984).

Where written and printed parts can be reconciled by any possible interpretation, however, such reconciliation must prevail. The law will not presume that parties intended to insert inconsistent provisions. In *Harrison v. New Brighton*, 110 A.D. 267, 97 N.Y.S. 246 (1st Dept. 1905), a dispute arose concerning a contract for construction of new sewers. Printed specifications provided that no tunneling would be allowed except by written permission of the village engineer. Typewritten provisions prohibited tunneling unless otherwise ordered by the engineer. The court found that there was no conflict between the typewritten and printed provisions. Reading the two provisions together simply indicated that the permission of the engineer to do tunneling must be in writing.

§ 2.11.6 General and Special Provisions

As you very well know, most construction contracts contain voluminous specifications with numerous provisions dealing with particular aspects of the construction. A question often arises regarding the priority of general versus special provisions. Where an item is dealt with both in general and in special provisions, it is the special provision that controls. This is particularly true where there is an inconsistency between the general provision and the special provision.

Two Guys from Harrison-N.Y., Inc. v. S.F.R. Realty Associates, 63 N.Y.2d 396, 482 N.Y.S.2d 465 (1984), involved a dispute over the terms of a long-term lease. The lease specifically allowed the removal of walls to permit access to improvements on adjoining property and authorized interior non-structural alterations. Since the lease specifically allowed these particular alterations to the leased building, a court will read the lease as implicitly prohibiting other alterations. The external structural changes proposed by the tenant would not be allowed.

§ 2.11.7 Multiple Contracts

Where several contracts are executed at the same time by identical parties and in the course of the same transaction, the law will treat those contracts as one. *Nau v. Vulcan Rail & Constr. Co.*, 286 N.Y. 188 (1941). The court will not, however, read two agreements together if one expressly supersedes the other. However, the general rule is that separate contracts relating to the same subject matter and executed simultaneously by the same parties can be treated as one contract. This is so even if one of the documents states that there are no other contracts between the parties and that the contract does not refer to the terms of other contracts. *Williams v. Mobil Oil Corp.*, 83 A.D.2d 434, 445 N.Y.S.2d 172 (2nd Dept. 1981).

§ 2.11.8 Matters Incorporated

An incorporation clause in a construction subcontract is that which incorporates by reference clauses in the prime contract between the owner and the general contractor. The incorporation clause will bind the subcontractor only to the prime contract provisions relating to the scope, quality, character and manner of the work to be performed. *S. Leo Harmonay, Inc. v. Binks Mfg. Co.*, 597 F.Supp. 1014 (S.D. N.Y. 1984), aff'd 762 F.2d 990 (2nd Cir. 1985).

Often, contract documents will specify the availability of additional reports, boring data, or surveys. Where the contractor is informed of the existence of this information, he or she is bound by the information in the reports. This is the case even if those reports are not physically attached to the bidding package. *Stuyvesant Dredging Co. v. U.S.*, 834 F.2d 1576 (Fed. Cir. 1987).

2.12 ENTIRE OR DIVISIBLE CONTRACTS

§ 2.12.1 Several Subjects

Whether a contract is entire or divisible is determined by reference to the intent of the parties. That intent is evident in an examination of the terms and provisions of the contract itself and the surrounding circumstances.

Pines of Islip, Inc. v. Island Concrete Corp., 22 Misc.2d 871, 196 N.Y.S.2d 252 (Nassau Co. 1959), involved a contract in which Island Concrete was to perform clearing, filling, grading, excavation, and cement work in connection with the construction of 250 houses. The delivery of fill was made optional with Island Concrete. Therefore, the court found that the portion of the contract for delivering of fill must be considered severable from the rest of the contract.

For the contractor, it is important to know that a lump sum construction contract is generally construed as being a single undivisible contract. *Tibbetts Contracting Corp. v. O&E Contracting Co.*, 15 N.Y.2d 324, 258 N.Y.S.2d 400 (1965), involved a contract with a developer. The contract was for a lump sum and was for excavation, grading, and drainage work. The lump sum contract was "single, entire and undivisible." The question of whether a contract is divisible bears directly on the amount to be collected as damages by a contractor when an owner breaches a contract. The owner breaching a contract will argue that the contract is made up of a number of clearly divisible components. Why is this?

Once a contract is breached, the contractor's compensation is determined by a court in one of two ways. The first allows compensation for the reasonable value of the work the contractor completed. On the other hand, the contractor may elect to proceed on the breached contract and collect the value of what he or she had lost. That amount is determined by taking the contract price, deducting payments made and deducted the cost of completion. By this method, the contractor's profit is preserved.

If no work had been performed nor any payments made by the owner, the contractor is entitled to the contract amount less the cost to complete the work, that is, his or her profit.

If the contract is deemed indivisible, the contractor's profit is figured on the entire contract price. An owner, of course, will argue that if the contract is divisible and, for example, that the contractor is only entitled to compensation for a profit for the first stage, say, clearing and grubbing, of the contract work.

The case of *New Era Homes Corp., Inc. v. Forster*, 299 N.Y. 303 (1949), illustrates this proposition.

New Era had contracted to make extensive alterations to Forster's home. The contract was for a lump sum. A certain amount was to be paid upon signing, an additional amount upon delivery of materials and start of work, a third amount upon completion of the rough carpentry and rough plumbing, and a fourth, and final, amount upon the completion of the work. The first two payments were made. Once the rough carpentry and plumbing were completed, New Era requested the third payment. Forster refused to make the payment and New Era stopped work on the project.

The court considered whether the contract was entire or divisible. If entire, there would be one price for performing the whole job, with payments on account at fixed points during the job progress. If divisible, the language of the agreement would establish several distinct and divisible obligations, each represented by an amount paid. The Court found that there was a total price for the whole of the work. Separately listed payments were not allocated absolutely to individual items of the work, but were merely scheduled part payments. Part payment for the work would be mutually convenient to both the contractor and to Forster. The court found nothing to suggest that the parties were making a separate agreement to perform a particular portion of the work for the amount of each partial payment.

The court felt it would be writing a new contract to break this promise into separate deals. The authority in New York and elsewhere indicates that a contract such as this expresses an intention that payment is conditioned and dependent upon completion of the work. The court cited both a New Jersey case, *Integrity Flooring v. Zandon Corp.*, 130 N.J.L. 244, and a Vermont case, *Peist v. Richmond*, 97 Vt. 97. This rule is reasonable. A person remodeling his or her home is committing to one plan and one result, not to a series of related projects. See also *Watson v. Auburn Iron Works, Inc.*, 23 Ill. App.3d 265, 318 N.E.2d 508 (Illinois);, *M.F.A. Mut. Ins. Co. v. Gulf Ins. Co.*, 445 S.W.2d 829 (Missouri); and *Fuller v. United Electric Co.*, 273 P.2d 136 (Nevada).

§ 2.12.2 Installment Contracts

An installment contract is one in which either performance (e.g., delivery of building supplies) or payment is made in independent and discrete installments. As demonstrated by *New Era Homes*, see above, an installment contract may not be divisible.

The courts will look to the intent of the parties. In *New Era Homes*, one of the parties attempted to associate certain payments with the performance portions of the work. The court found that contract to be entire and not divisible. In the ordinary installment contract, delivery may be made on request for payments made periodically over time. The courts will find a contract of this nature to be entire and nondivisible.

For the performance of a contract to be deemed divisible, the courts would want to see, for example, delivery of different items at different prices. The argument for divisibility would be aided if a payment was due immediately upon receipt of those items.

The case law strongly suggests, however, that most construction contracts are considered entire rather than divisible. Because of the size and cost of construction contracts, there is a long tradition of what the industry calls *progress payments*. These could qualify as installment payments; part payment for part performance. However, the fact that the industry considers them progress payments indicates the contracts are considered indivisible and entire. The progress payment is merely a vehicle that allows the contractor to get on with the constructing of the project and leaves the financing to its owner.

Although payment is an important consideration, it is the intent of the parties which is paramount when considering whether or not a contract is divisible. *Peck v. Ives*, 84 N.M. 62, 499 P.2d 684 (New Mexico); *Allen-Howe Specialties Corp. v. U.S. Constr., Inc.*, 611 P.2d 705 (Utah, 1980).

§ 2.12.3 Progress Payments

Progress payments are a reality in the construction industry. You, as a contractor, cannot be expected to finance the often stratospheric costs of a construction contract while awaiting for payment of the lump sum amount at the end of the project. Are then construction contracts considered divisible? Hardly ever.

A contract for construction of a swimming pool was determined to be entire and not divisible. The contract called for payment of a portion of the contract amount at completion of the pool shell, with the remainder to be paid at completion of the pool. The court looked to the intent of the parties. The record was devoid of any evidence that the parties considered this to be two separate agreements. Payment of one progress payment would not settle in full the obligations under that portion of the contract. *S. & W. Investment Co. v. Otis W. Sharp & Son, Inc.*, 247 La. 158, 170 So.2d 360 (Louisiana).

In *Shalman v. Bd. of Ed.*, 31 A.D.2d 338, 297 N.Y.S.2d 1000 (3rd Dept. 1969), the court looked to the intentions of the parties to the contract. This was a lump sum contract to perform certain electrical installation work at designated school buildings. The court stated that, in general, the inclusion of a provision for progress payments does not make a construction contract divisible. The intent of the parties is a primary factor in determining whether a building or construction contract is entire or divisible.

2.13 EXTRA WORK

§ 2.13.1 Modification

The procedure for modifications to a contract is generally outlined in one of the contractual provisions. That provision should be read closely. The owner will often require that payment for modification is contingent upon receipt by you of a written directive from the owner or his or her architect. In the alternative, if the modification to the contract has been suggested by you, the owner will require prior written agreement by him or her to the modification. Often, the modification clause contains requirements for filing a notice of claim (see § 2.14.7) within varying time periods. In the flurry of activity during the bidding process, these provisions are often, and rightly so, overlooked. They are certainly nonnegotiable. However, once you receive the award of a contract, these provisions should be carefully studied and protocols established within your company to insure that the provisions are adhered to strictly. If they are not, the courts will reject your claims, particularly if those claims are against a public agency.

In *Buckley v. City* 121 A.D.2d 933, 505 N.Y.S.2d 140 (1st Dept. 1986), appeal dismissed 69 N.Y.2d 742, 512 N.Y.S.2d 1030 (1987), a pumping station contractor sought to recover damages in connection with the redesign of a cofferdam designed by the City to prevent seepage into the excavation. The court discussed the often tenuous distinction between "extra" and "additional" work. The contract defined "extra" work as work other than that required by the contract at the time of its execution. The court felt that this definition would seem to include "additional" work, which is caused by unanticipated subsurface conditions. However the work is categorized, the court felt that it is compensable only by means of a modification pursuant to the contract.

The contract provisions provided that work can only fall into one of three categories. First, work whose performance is required by the contract allows for no additional compensation. Second, for extra or disputed work, the contract requires strict compliance with the contractual provisions. Absent such compliance, claims relating to the extra or disputed work are explicitly waived. Third, if the work involved is additional work, attributable to a changed condition, there must be issued a contract modification subject to the Commissioner's written approval, as well as prior written approval of the Comptroller's chief engineer for any increase in cost.

Buckley argued that the City's resident engineer had waived strict compliance with the contract procedures.

Unfortunately, one of the articles of the contract provided that the resident engineer had no power to change the contract terms. The court found that Buckley's failure to seek a modification regarding the extra work was fatal to its claim.

Howdy Jones Const. Co., Inc. v. Parklaw Realty, Inc., 76 A.D.2d 1018, 429 N.Y.S.2d 768 (3rd Dept. 1980), appeal denied 51 N.Y.2d 836, 433 N.Y.S.2d 759 (1980); aff'd 53 N.Y.2d 718, 439 N.Y.S.2d 354 (1981), involved a contract to excavate and landscape a racetrack. The trial court disallowed invoices for extra work finding that they were not signed by personnel of Parklaw authorized to approve extra work. However, witnesses of Howdy Jones testified that the invoices represented work done by them. The work had not been contemplated in the original contract. Parklaw had requested that the work be done. The testimony illustrated a course of conduct in which Parklaw personnel did not sign invoices until after extra work was performed. The appellate court found for Howdy Jones. Extra work outside the scope of a contract is recoverable, even if the request is not in writing. If directed by the owner, the basis for payment will be an independent oral contract.

A claim for extra work cannot be sustained if the work in question was included within the original contract, plans, and specifications. The proposition is illustrated in the case *John Monks & Sons v. West Street Improv. Co.*, 149 A.D. 504, 134 N.Y.S. 39 (1st Dept. 1912). One of the multiple defendants in that case was John Pierce Company, a contractor engaged to erect a building for another defendant, West Street Improvement Company. Pierce solicited bids from a number of excavation and foundation subcontractors. Monks submitted a proposal in writing to excavate the entire area of the lot down to a level of 13 feet, 11 inches below the curb, to furnish and drive piles as shown on the plans, to excavate for and build concrete piers, to shore and underpin adjoining buildings, to excavate for a boiler pit, to sheet pile the streets, to build sidewalk bridges, and to do all necessary pumping for a lump sum. In addition, Monks was to drive extra piles, make extra excavation, and furnish materials for mixing and placing concrete for the cellar floor at stated unit prices. Pierce accepted the proposal upon the condition that a satisfactory agreement should be signed. Without waiting for this agreement to be made or a formal contract executed, Monks commenced to excavate for the building. Pierce prepared and submitted to Monks specifications and a formal contract. Monks refused to execute the contract. In the progress of its work, it discovered rock-filled cribbing where the piles were to be driven. The condition differed from that indicated on the blueprints. Pierce wrote to Monks and suggested that it complete the work and if Monks was correct, he could recover extra compensation. The court found that the sketch of test borings and the blueprint showed what anyone familiar with the location would have known; i.e., that the area to be excavated was filled in or made land upon river mud and sand. The court found that the test borings and blueprints do not purport to be complete representations of materials to be excavated or to be encountered in the driving of piles. The court found that in view of Monks' familiarity with the site, his presumed knowledge and probable expertise, he should have expected that stone-filled cribs, as well as earth-filled cribs would be encountered. It was incumbent upon Monks to investigate and ascertain the situation for himself. The proposal did not show that Monks contracting upon the assumption that only earth-filled cribs would be encountered.

The court found that it would be manifestly unjust to pay Monks for work that was included in the original contract. The driving of the piles through stone cribbing did cost Monks more than what driving through earth-filled cribbing would have cost. However, Monks made a bad bargain and must abide with that bargain. See *Watson Lumber Co. v. Guennewig*, 79 Ill. App.2d 337, 226 N.E.2d 270 (Illinois). He cannot be compensated for work included in the original contract.

Just what is included in the contract and how much can a court presume? In *Murphy v. No. 1 Wall Street Corp.*, 142 A.D. 835, 127 N.Y.S. 735 (1st Dept. 1911), a subcontractor made an extra work claim for grinding brick at the corner of a building. The building was erected at the intersection of Wall Street and Broadway in New York City. That intersection is not a right angle. In order to construct the building in a workmanlike fashion, it was necessary to bevel a number of bricks at the corner of the building. The court found that the subcontractor knew when he took the contract that the corner of the building would not form a right angle and that he could not perform his work without beveling and grinding the corner bricks. The subcontractor must be deemed to have included in this beveling and grinding in his initial estimate. The brickwork subcontractor testified that a foreman for the general contractor verbally authorized the grinding of the brick. The court finds that this does not amount to anything more than approval of the method of beveling. The grinding and beveling was work included in the original subcontract and, therefore, must be disallowed as an extra work claim.

Finally, not only does the extra work provision in the contract have to be carefully read, but the contract documents as a whole must be considered. Is there a conflict between the provisions of the contract and the specifications? In a contract for public work, the provisions of the contract itself must govern. *McGarry Contracting Co. v. Bd. of Ed.*, 284 N.Y. 218 (1940), concerned a claim against the Board of Education for extra work in connection with subsurface conditions. In the progress of its excavation, McGarry encountered foundation walls, rock and other solid substances. The Board of Education modified its plans and specifications to provide that footings and foundation piers be installed at a lower elevation than originally specified. Without considering the nature of the addition work involved in removing the foundation walls and rock, the Board merely paid the contractor on a unit price basis for a volume of material removed, as was specified in the contract specifications. The court pointed out that the contract document itself provided that should extra work be involved, the Board of Education was responsible for the reasonable value of that work. The reasonable value would be added to the contract price. The Board of Education attempted to subvert the provision of its own contract by arguing that removal of the old walls was covered by the unit price provision of the specifications. The court rejected that argument and held that the conflict between the contract and its specifications require that the contract document control.

What other contract provisions may affect a claim for extra work?

In many contracts, there is a provision relating to site visits by the contractor prior to bid. In such provision, the contractor agrees that he or she has reviewed contract documents, reviewed the site and surrounding territory, and general fully informed himself or herself as to the labor and materials to be furnished. Often, the provision also states that the contractor has secured this information from his or her own personal investigation and not from estimates, records, files, or documents of the owner. Where such a provision exists, your claim for extra work is that much more difficult. The law implies your knowledge of the soil and of subsurface conditions.

Savin Bros. Inc. v. State, 62 A.D.2d 511, 405 N.Y.S.2d 516 (4th Dept. 1978), aff'd 47 N.Y.2d 934, 419 N.Y.S.2d 969 (1979), involved a claim for the use of material excavated from the site in embankment and backfill construction. The contract specified that as Savin excavated material, that material was to be used in the construction of various embankments. The contract explicitly stated that Savin was to absorb the costs for correcting high moisture content in the soil. In order to insure that the embankments were properly constructed, the embankment was subject to a proof rolling test.

Savin proceeded to excavate various areas. When the material excavated was placed in the embankment and proof tested, it failed the test. The State refused to permit Savin to re-excavate the top two feet of the embankment and replace the unsuitable material with select fill. However, Savin proceeded to do so to remedy the situation and progress the project. The court held Savin liable for the reexcavation and placement of the select barrow.

It is a general rule that recovery cannot be had for extra work when that work falls within the contract plans and specifications.

Here, all the experts, including Savin's, agreed that the material originally placed in the embankments would have been suitable if properly treated. In addition, the contract contained a clause regarding inspection of the site by the contractor.

The court found that rather than extra work, this was work within the scope of the contract. Savin was obligated to treat the embankment fill. Any cost Savin incurred by removing the embankment fill and replacing it with select barrow was not compensable as extra work. This was merely extra cost incurred to perform the work included in the contract and, therefore, Savin was liable for this cost.

Is the term "suitable material" subject to dispute? Of course. In *Frank A. Scibetta Plumbing & Heating Corp. v. M & W Ltd. Partnership*, 90 A.D.2d 956, 456 N.Y.S.2d 544 (4th Dept. 1982), aff'd 58 N.Y.2d 1092, 462 N.Y.S.2d 848 (1983), a contract provided that excavated areas were to be backfilled with six inches of clean, tamped sand over piping and then suitable material to grade. After laying exterior water and sewer pipes, Scibetta placed sand over the pipes and backfilled the trenches with excavated earth. The project owner objected and insisted that the contract required select or porous fill rather than excavated earth. Scibetta complied with the owner's direction and claimed entitlement for the cost of this work.

Although Scibetta was required to use "suitable material," that term was not defined in the contract or in the specifications. Perhaps the owner will do well to look to his architect for reimbursement for this cost since the court found for Scibetta. Although a section of the contract provided that a porous fill was to be used for certain work, that section specifically excluded backfilling for the water and sewer pipes. In addition, there was no testimony that the owner's architect had made any determination as to the suitability of the excavated earth placed in the trenches.

Although it is important to be aware of the contractual provisions regarding claims for extra work, in their attempt to do justice, the courts will enlist the aid of the doctrine of waiver. Although a contract may specifically lay out the procedure necessary to pursue a claim successfully for an extra, the parties, by their acts, may ignore the applicability of the clause and thereby waive its effect. In *Sturdy Concrete Corp. v. Nab Const. Corp.*, 65 A.D.2d 262, 411 N.Y.S.2d 637 (2nd Dept. 1978), appeal dismissed 46 N.Y.2d 938, 415 N.Y.S.2d 212 (1979), appeal after remand 89 A.D.2d 588, 452 N.Y.S.2d 252 (2nd Dept. 1982), Nab was awarded a contract by the Port Authority of New York and New Jersey for reconstruction of the roofs at North River Piers 88, 90, and 92. By subcontract with Nab, Sturdy was to perform certain specified concrete work. Under the subcontract, extra work was required to be subject to a separate written agreement. From the evidence, the court found that, except in one instance, the work was not ordered in writing. Furthermore, there were no facts that could demonstrate that Nab waived the applicability of this requirement. Although Sturdy's project manager testified that Nab had authorized the work orally, that contention was disputed by Nab's witnesses. The issue of the witnesses' credibility was decided against Sturdy by the trial court. The appellate court would not disturb that finding. Sturdy did not receive compensation for the alleged extras.

In a dispute over work during the performance of the contract, may the contractor immediately proceed to court? The real issue is whether the contract contains an extra work provision.*Kalisch-Jarcho, Inc. v. City of New York*, 72 N.Y.2d 727, 536 N.Y.S.2d 419 (1988), involved a dispute between a plumbing contractor and the City of New York in connection with construction of a new broom depot and salt storage building for the Department of Sanitation. On the day of the contract signing, before the contract was signed, a dispute arose regarding concrete pads to be installed below underground fuel tanks. Kalisch-Jarcho was clearly required to install tanks and concrete casing around them. However, Kalisch-Jarcho felt that the excavation for and installation of the concrete pads below the tanks were the responsibility of the general construction contractor. Pursuant to the terms of the contract, the dispute was submitted to the Commissioner of the Department of Sanitation for resolution. The Commissioner determined Kalisch-Jarcho was to excavate for and install the concrete pads below the tanks.

Kalisch-Jarcho then proceeded to court to request a declaratory judgment. In a declaratory judgment action, a court will declare the particular rights of a party. Kalisch-Jarcho sought to have a declaration that it was not contractually obligated to perform the disputed work. Using a novel theory, Kalisch-Jarcho argued that if it performed work totally outside of the scope of the contract, it would not be entitled to compensation even if performed under protest. The legal theory Kalisch-Jarcho raised was established to protect a municipality against conspiracies between a contractor and a municipal employee to obtain payment for work not contemplated by the contract.

The court found that parties to a contract may not seek a declaration of contractual rights when the contract specifies a different and reasonable means for resolving the contractual dispute. Here, the excavation and installation of the concrete pads below the underground fuel oil tanks was shown on the contract drawings and was a necessary part of the project. There was no question of subjecting the municipality for unnecessary, unanticipated costs; nor was or could it be alleged that there existed collusion between the contractor and a municipal employee. The dispute here did not threaten injury to the public interest. Kalisch-Jarcho was required to perform the work pursuant to the terms of the contract.

How responsible is a contractor for his or her failure to act? Completely responsible. *Delma Engineering Corp. v. 6465 Realty Co.*, 39 A.D.2d 846, 332 N.Y.S.2d 841 (1st Dept. 1972), aff'd 31 N.Y.2d 816, 339 N.Y.S.2d 464 (1972), involved an action by Delma against 6465 Realty for work on an apartment building then under construction. Delma agreed to perform the excavation and foundation work. The contract also required that Delma was to make a survey of the work site. This Delma did not do. Subsequently, public utilities structures were unexpectedly discovered within the excavation lines. Delay was incurred when it was necessary to perform replanning and relocation. Pursuant to its contractual obligations, Delma was to protect such installations.

The court found that due to Delma's failure in completing the survey, the extra costs incurred should not be paid for by the owner. Delma, by contract, was to perform a particular and specifically delineated job at an agreed price. The fact that it could not do so was owing to its own failure to perform its survey work. Delma was not entitled to extra work where such work could have been avoided by a prudent examination and survey of the site.

In *Kenison v. Baldwin*, 351 P.2d 307 (Oklahoma 1960), Baldwin, a general contractor, entered into a written contact to build a home for Kenison. Kenison orally requested Baldwin to make certain changes in the plans, accepted the home, and moved into it. Although the contract stated that changes in the plans and specifications were to be in writing, the circumstance indicated the existence of an independent oral agreement. The parties were bound by that agreement, and Kenison was bound to pay Baldwin.

On the other hand, courts have held to the contrary. In *Morris Glasser & Sons, Inc. v. Jonwal Constr. Co.*, 186 Misc. 253, 60 N.Y.S.2d 25 (App.T. 1946), the court noted that the agreement in question expressly provided that a claim could not be made unless it is in writing and signed. The court held any alleged oral modification is ineffective.

While an independent oral contract, waiver, or modification may be recognized by the court to circumvent a provision requiring written authorization, the facts need to be clear. See *Harrington v. McCarthy*, 91 Idaho 307, 420 P.2d 790; *Hoffman v. Glock*, 20 Md. App. 284, 315 A.2d 551 (Maryland). Furthermore, if the owner has benefited from the extra work, courts may imply an agreement to pay reasonable compensation for such work.*Black Lake Pipe Line Co. v. Union Constr. Co.*, 520 S.W.2d 486 (Texas Civ. App.).

§ 2.13.2 Required by Engineer or Architect

When a contract allows an engineer or architect to determine the manner of performing a contract, can the contractor receive compensation for extra work? The answer is clearly no. It is important to recognize, however, that what is being discussed here is not so much a purely construction contract, but rather a performance contract. The seminal case in this area is *Snare & Triest Co. v. New York*, 191

A.D. 184, 181 N.Y.S. 304 (1st Dept. 1920), aff'd 233 N.Y. 528 (1922). The contract involved strengthening the end spans of the Williamsburg Bridge over the East River. Snare & Triest asserted a claim for numerous extras. Snare & Triest alleged that they could not follow the plans and specifications strictly, nor were they permitted to do so, but were directed by the engineer to perform certain extra work.

Pursuant to the contract, the engineer was given the authority to determine the amount or quality of the kinds of work to be performed under the contract. The engineer was further given the authority to reject, at any stage of the work, any material or workmanship that he felt was unfit. The court noted that the job was experimental at best. It was a more technical project than building a new bridge and involved replacing and adding to various units in order to strengthen an existing structure.

The court looked to the contract. Was it to definitely strengthen the end spans of the bridge or was it to do certain work in a particular manner in accordance with specific plans? The court felt that if the contract was to accomplish the definite purpose of strengthening the end spans of the bridge, then all work performed at the direction of the engineer was work and material incidental to performance of the contract. The contractor could not recover for that work, as it was within the scope of the contract.

If the contract was to perform specific work and furnish specific materials in accordance with the plans, then the claimed extra work was entirely outside the scope of the contract and also not subject to recovery by plaintiff.

This is a harsh result, yet as the court saw it, Snare & Triest had signed a contract that allowed the engineer to determine its scope during performance. The court's alternative reasoning, that if the contract was to perform specifically, the direction of the engineer was totally without the scope of the contract, is a position that has slowly eroded over the years. The contract did contain a provision for notifying the commissioner of claims for additional work or materials. Snare & Triest did not procure contractually required authorizations of the commissioner for the performance of the extra work. Ignoring a theory of waiver, the court states that Snare & Triest is foreclosed from enforcing its claims.

Performance contracts cry out for review by your attorney before contract signing.

A more recent case demonstrates the courts' increasing sophistication in dealing with construction contracts. In *Thomas Crimmins Contracting Co., Inc. v. New York*, 138 A.D.2d 138, 530 N.Y.S.2d 779 (1st Dept. 1988), aff'd ctfd. ques. ans., on banc, 74 N.Y. 166, 544 N.Y.S.2d 580 (1989), Crimmins, a subway construction contractor, sued for extra work. The court defined extra work as work necessarily required in the performance of the contract arising from circumstances not anticipated at its signing. The City argued that pursuant to an alternative dispute resolution clause in the contract, once the contractor had submitted its claims to the Transit Authority's chief engineer, his decision was final and binding. The court held that the engineer's authority did not extend to deciding questions of law or issues of contract construction. The court felt that to give the engineer, who was obligated to answer to his employer, the City, the power to conclusively bind the contractor on legal determinations is to expose the contractor to undue risks of performing without compensation. This would be tantamount to conferring upon the engineer the unilateral power to modify the agreement, imposing risks on the contractor, which he did not assume. A municipality cannot be final and conclusive arbitrator over its own disputes.

§ 2.13.3 Where Caused by Another

M.L. Ryder Bldg. Co. v. Albany, 187 A.D. 868, 176 N.Y.S. 456 (3rd Dept. 1919), upholds the rule that where parties enter into a contract, there is an implicit obligation that the person for whom the work is done will not obstruct or hinder the performance of that work. Ryder entered into a contract to erect a one-and-a-half-story fire alarm signal station. Upon completion of the contract, it sued for extra work, necessitated by reason of the acts, directions, and omissions of the City.

The City's contract contained a provision which would not allow extras unless they were submitted in writing to the City's architects and approved.

The court found that where a municipality, by its own act, causes work done by a contractor to be more expensive than it would otherwise have been, the City is liable for that increased cost or extra work.

The courts will afford a contractor only so much leeway, however. In *E.M. Substructures Inc. v. City of New York (Sewers 149th St., Queens)*, 73 A.D.2d 608, 422 N.Y.S.2d 444 (2nd Dept. 1979); appeal dismissed 49 N.Y.2d 878, 427 N.Y.S.2d 990 (1980), E.M. sought to recover for extra work. However, prior to the lawsuit, E.M., in exchange for an extension of time to complete the contract, had signed a waiver of claims without exception. E.M. did not dispute that it could have preserved its claims by excepting them from the waiver. Not only did E.M. concede that it was aware of the claim for extra work when it signed the waiver, but also stated that on past occasions, it had preserved such extra work claims by excepting them from waivers.

This is just the sort of situation where an attorney should be present with you. The claims for extra work were rejected by the court.

The liability for delay between the general contractor and his subcontractor is governed by the general rule that a contractor is not responsible for delays suffered by his subcontractor unless they are caused by the prime contractor or that the contractor has made a contractual commitment to be responsible for those delays. See *McGrath v. Electrical Construction Co.*, 230 Or. 295, 364 P.2d 604 (Oregon); *Doyle & Russell Inc. v. Welch Pile Driving Corp.*, 213 Va. 698, 194 S.E.2d 719 (Virginia).

§ 2.13.4 Necessitated by Casualty

Casualty or unforseen difficulty in performing the work does not entitle a contractor to an increase in his or her lump sum price. Where a contractor has bargained to perform certain work for a certain price, the courts will hold him or her to that bargain.

Leahy v. Lucius Engineering Co., 186 A.D. 354, 174 N.Y.S. 310 (1st Dept. 1919), aff'd 227 N.Y. 660 (1920), involved the construction of an elevated portion of the New York City subway system. Leahy had contracted with the Interborough Rapid Transit Company. He then subcontracted with Lucius for Lucius to receive, haul, load, distribute, erect, and rivet steel, and to apply two coats of paint. By reason of the nondelivery of steel, Lucius was unable to commence work for over a year. Lucius wrote to Leahy, advising that unless a new agreement was reached, it could not continue to work, due to the increased costs in labor and materials. Leahy notified Lucius that unless it proceeded with the contract, it would treat its refusal as a breach and hold Lucius liable for any damages. The court found that Lucius breached the contract. Lucius had no right to demand a modification of the contract. A change in price, either in labor or materials, is one of the risks of doing business.

The court admitted that this was a hard case. If the steel had been promptly furnished, Lucius presumably would have made a profit. In addition, Lucius was not responsible for the delay. Undoubtedly, if Lucius had, by notice, put Leahy in default, Lucius could have terminated the contract or could have performed and recovered the damages suffered by reason of the delay. The option would have been readily apparent to an attorney, but was not to Lucius.

Lucius was held responsible for Leahy's increased costs.

2.14 WRITTEN ORDER FOR EXTRAS

§ 2.14.1 Application

The general rule upholds a contractual provision that requires that orders for extra work be in writing. Such a provision can only be circumvented where it is shown that there has been a waiver of such provision by the parties to the contract. When does waiver of a contractual provision arise? Clearly not in *Bibeau Constr. Co. v. Hauser Bros., Inc.*, 39 A.D.2d 955, 333 N.Y.S.2d 459 (2nd Dept. 1972), aff'd 33 N.Y.2d 883, 352 N.Y.S.2d 445 (1973). Bibeau, a subcontractor for excavation and installation of underground utilities, sued the plumbing contractor. Bibeau claimed for extra excavation it performed when another contractor on the project failed to properly grade the job site in accordance with the construction plans and specifications. Bibeau performed the excavation even after Hauser Bros. had written to Bibeau and stated that Bibeau should do no extra work on the project unless Bibeau received a written order from Hauser. The letter continued: "If you do, it will be your responsibility, not ours." The court held that Bibeau could not collect against Hauser for the purported extra excavation. See also *Comet Heating & Cooling Co. v. Modular Technics Corp.*, 57 A.D.2d 526, 393 N.Y.S.2d 573 (1st Dept. 1977).

Where it was the undisputed custom on the job that written authorization were obtained for extra work, the contractor would not be entitled to a claimed extra absent a writing. *Valente v. Two Guys from Harrison, N.Y., Inc.*, 35 A.D.2d 862, 315 N.Y.S.2d 220 (3rd Dept. 1970). See also *Piper v. Murray*, 43 Mont. 230, 115 P. 669

(Montana 1911). However, where the owner's agent refused to sign written change orders, as required by the contract, but assured the contractor that such orders need not be in writing, the contractor will be allowed to present his or her proof regarding the waiver of the written change order provision. *Wagner v. Graziano Constr. Co.*, 390 Pa. 445, 136 A.2d 82 (Pennsylvania 1957).

In a dispute between a general contractor and his site demolition subcontractor, in connection with the construction of a municipal parking garage, extra compensation was properly awarded to the subcontractor, even where extra work needed to be approved in advance and in writing. The facts show that those provisions had been disregarded by the parties and that, either orally or by implication, the general contractor had waived the writing requirement of the extra work provision. *D.M. Holden, Inc. v. Contractor's Crane Service, Inc.*, 435 A.2d 529 (New Hampshire 1981).

The contract provision was clearly waived where the owner, a former contractor and carpenter, was present on the job, acquiesced to changes, made partial payments, and even approved of the work. *Swenson v. Lowe*, 5 Wash. App. 186, 486 P.2d 1120 (Washington).

In *Hedden Constr. Co. v. Rossiter Realty Co.*, 136 A.D. 601, 121 N.Y.S. 64 (1st Dept. 1910), aff'd 202 N.Y. 522 (1911), the contractor began work on a reinforced concrete garage. Part-way through the work, new specifications were issued requiring work not in the original contract specifications. The new specifications included a provision that extra work must be ordered in writing. The court found that the revised specifications, requiring a writing for extra work, only applied to work to be performed in the future.

The owner was required to pay for the work in the revised specifications, which was additional to the work in the original specifications.

Derrico v. Muller, 142 N.Y.S. 479 (App.T. 1913), involved a saloon owner who hired Muller to make certain repairs to his property. It was Derrico's contention that the contract provided that no extra charges should be paid for or allowed beyond the amount mentioned in the contract. The court found that the contract could be modified between the parties. The saloon owner also argued that any allowance for extras had been required by the contract to be in writing. The court found that this requirement could also be modified by the parties.

§ 2.14.2 Definition

What field conditions will justify extra work? Where the contractor should have known of the existence of certain physical characteristics, as in *Murphy* (§ 2.13.1) or should have performed certain work and did not, as in *Delma* (§ 2.13.1) the courts will reject compensation for any additional costs incurred.

Often, the proponent of a contract will attempt to define extra work in its provisions. The federal government attempts to define differing site conditions in Part 52 of the Federal Acquisition Regulation. That provision requires written notification to the contracting officer with respect to subsurface or latent physical conditions at the site differing materially from those indicated on the contract, or unknown physical conditions of an unusual nature. The American Institute of Architects, in their document A-201, ¶ 4.3.6, also provide a protocol requiring notice for claims for concealed or unknown conditions.

In *American Pipe & Constr. Co. v. State*, 247 N.Y. 150 (1928), American Pipe had contracted for construction of a portion of the New York State Barge Canal. The contractor installed certain sheeting and bracing to hold back soft materials along the canal embankment. The court found that the contract itself required the erection of sheeting and bracing to confine this material. The court found it doubtful whether the engineer's order should have been sought, since in order to comply with the contract, the contractor must have installed the sheeting and bracing.

Where items of extra work are provided, pursuant to a verbal arrangement, the court will deem that circumstance to be the execution of an oral contract, to which the parties are bound. Although the written contract may call for written change orders for extras, a contractor is obligated to his or her subcontractor for the cost of additional lighting fixtures, notwithstanding the absence of a writing. *Moore v. Continental Casualty Co.*, 353 F.Supp. 105 (Oklahoma).

So too in *Roff v. Southern Constr. Corp.*, 163 So.2d 112 (La. App., 1964), which involved a contract between a general contractor and a subcontractor to lay and caulk bricks in connection with the construction of homes. Although the terms of the contract required a written order for extra work, the court found for the subcontractor and awarded extra costs for removing dirt found on top of the foundations of the houses on which the brick work was to be done. The court reiterated the general rule that claims for extra work would not be allowed unless written authority was given pursuant to a contract provision. However, the court found here that the general contractor was aware of the extra work, had not objected to the subcontractor's performance of the extra work, and that the general contractor's superintendent had directed the subcontractor to proceed with the work and authorized him to charge the general contractor. Since the extra work was necessary, the contractor was aware of the extra work, and there was a subsequent oral agreement authorizing the work, Louisiana courts allowed payment to the subcontractor.

§ 2.14.3 Buildings

F.W. Carlin Const. Co. v. New York & Brooklyn Brewing Co., 149 A.D. 919, 134 N.Y.S. 493 (2nd Dept. 1912), involved a building contract that contained a provisions requiring a written order for any extra work. The contractor argued that, in bad faith, both the owner and his architect increased the work under the contract's drawings and specifications, required him to do things not required under the contract, and then refused to give written change orders for that work. The architect, collusively with the owner, refused to certify the items claimed as extra work. The court found that the conduct of the owner and his architect did not in any way suggest renunciation of the contract. On the contrary, the record indicates that the owner and its architect insisted on all their original rights. These facts do not indicate a waiver of the contract provision but, on the contrary, an attitude of refusing to give up the rights to which the owner was entitled.

The contract in *Hi-Valley Constructors, Inc. v. Heyser*, 163 Colo. 1, 428 P.2d 354 (Colorado), contained a provision requiring written authority for changes. Construction plans were incomplete, and the homeowners visited the site almost daily, periodically ordering extras and changes. Although both the contractor and the homeowners realized that these changes would cost more money, no written direction was ever obtained for the changes. The court found that both parties, by their actions, had waived the provisions requiring written authorization.

§ 2.14.4 Necessity of Writing

From the cases above, it is obvious that the courts will attempt to enforce a provision requiring written authorization for extra work. However, where such authorization is not forthcoming, the courts will often seek to do justice. If the owner, or contractor in the case of a subcontractor's claim, was aware of the changes, and the changes were necessary for the work, courts will be inclined to see waiver and overlook the written authorization provision. Additionally, where the owner or his or her authorized agent directs the contractor to perform extra work, the courts will find the existence of an oral contract independent of the written contract. This is the case even where the written contract specifically requires that any change orders be made in writing. Once the contractor has provided the work requested by the owner, the courts will consider the oral contract executed and require the owner to reimburse the contractor for these extras. However, where the parties' conduct or course of dealings do not indicate a waiver of the provisions requiring written authorization, or where, for example, all other change orders are in writing, the courts will uphold the contractual provision and disallow extra claims that are not the subject of a writing.

§ 2.14.5 Authority

What comprises a written order for extra work? In *Fetterolf v. S. & L. Const. Co.*, 175 A.D. 177, 161 N.Y.S. 549 (2nd Dept. 1916), a contract provided that no alternations could be made except upon the written order of the architect. The architect furnished working plans or sketches to the contractor that indicated changes enlarging the attics in a number of houses. The court held that the working plans or sketches could not be regarded as a written order of the architect within the meaning of the contract provision. This is unfortunate news to the contractor since it is often the architect with whom the contractor has sole contact during the construction process. As discussed in § 2.14.8, the contractor must look very closely at the authorizations provided to the architect. In general, changes can only be approved by the owner or the owner's agent. The courts have often reviewed the authority given to an architect or engineer and found that authority insufficient to authorize changes pursuant to the written change order provisions of the contract.

§ 2.14.6 Waiver of Provision

The general rule is that in spite of a contract provision that extra work must be supported by a written authorization, the courts will attempt to do justice. Where an owner knowingly receives and accepts the benefits of extra work, which is directed by the owner or by his or her agents, there is a waiver of contractual requirement for a writing. *S. Leo Harmonay, Inc. v. Binks Mfg. Co.*, 597 F.Supp. 1014 (S.D.N.Y. 1984), aff'd 762 F.2d 990 (2nd Cir. 1985).

In *Care Systems, Inc. v. Laramee*, 155 A.D.2d 770, 547 N.Y.S.2d 471 (3rd Dept. 1982), Care Systems was to construct an addition to Laramee's home. Care Systems stated that Laramee directed that additional, corrective work be performed when he was informed by Care Systems of an improperly laid foundation. The record is silent as to whether Laramee denied this direction. Rather, Laramee relied on a provision in the contract requiring a writing for an change or modification. The court found that where an owner knowingly receives and accepts the benefits of extra work, orally directed by him or her or his or her agents, then the owner is bound to pay the reasonable value of that work. This is the case notwithstanding contract provisions that extra work must be supported by a written authorization, signed by the owner. The owner's action waives the requirement for a writing.

§ 2.14.7 Notice of Claim

The notice of claim requirement, often found in statutory law, applies to various governmental agencies. In New York State, there is a requirement for a notice of claim with respect to claims against boards of education, various agencies, towns, villages, cities and the State itself. The requirement for a notice of claim differs from the requirement for a written authorization for change work in one important aspect. Since the notice of claim requirement is found in the law itself, the courts have consistently held that the filing of a notice of claim is a condition precedent to recovery.

We met condition precedents in § 2.6.2. A condition precedent must occur, if recovery is to be permitted.

The operation of a notice of claim statute is illustrated in *G.A. Contractors, Inc. v. Bd. of Ed.*, 176 A.D.2d 856, 575 N.Y.S.2d 334 (2nd Dept. 1991). G.A. was to perform certain construction work at a school. It performed extra work and claimed it was entitled for payment for that work. The Board of Education countered, by a document dated January 30, 1987, and stated that no change order had been issued and that the work had been unauthorized. G.A. filed its notice of claim on February 2, 1987. It was the Board's contention that the claim accrued either on April 17, 1986, when G.A.'s president wrote to the Board indicating that the job was 98% complete, or in September 1986, when G.A. submitted its last payment application to the Board. If either of those dates were deemed to be the date upon which G.A.'s damages became ascertainable, the notice of claim would be filed too late. The applicable statute requires that the notice of claim be filed three months after accrual of the claim. The appellate court held that the trial court should determine when the claim accrued and should accord appropriate weight to any evidence that indicated when the extra work was substantially complete or when a detailed bill for the extra work was submitted by G.A. to the Board.

An unusually harsh application of the notice of claim requirement is illustrated in *Prote Contracting Co., Inc. v. Bd. of Ed.*, 567 N.Y.S.2d 687 (1st Dept. 1991). A school contractor was asserting a claim for extra work. The court dismissed the claim, noting that a notice of claim was not filed within the statutory three months "after the accrual of such claim." The facts of the case, however, demonstrate unusual harshness.

This claim involved extra work incorporated in a change order approved by the Board of Education. Not only did the Board of Education approve the change order, but the Board accepted a payment requisition, which included the amount of the change order. Sixteen months after the submission of the payment requisition, Prote Contracting was forced to litigate in an attempt to receive payment for the approved change order, which had appeared on the accepted payment requisition. The court held that the claim for the extra work accrued upon the completion of the work. Prote concededly allowed 16 months to pass before filing its notice of claim immediately prior to litigation. The court found that the condition precedent was not met and dismissed Prote's claim.

The record does not indicate whether the court considered the approved change order as containing the necessary statutory requirements of a notice of claim. Since the change order is generally signed by a contractor, it would seem that the requirement, such as the scope of the extra work, the materials employed, and the amount claimed, should have been contained in the change order. The fact remains: the courts strictly interpret and enforce conditions precedent. The time by which to file a notice of claim should be strictly monitored by the contractor. If payment is not received, even on an approved change order, a notice of claim should be filed.

This is an area rife with danger to the contractor. The claim in *Prote* was almost $90,000. When you contract with any public entity, you should immediately speak with your attorney and determine if any statutory conditions precedent exist. As stated earlier, in New York, notice of claim provisions are applicable to contracts with boards of education, various agencies, villages, towns, cities, and the State. Other states have similar conditions precedent. Other notice provisions may be contained in the contract itself. In addition, your attorney should review any contract with private entities, for contractual conditions precedent, such as those regarding payment (§ 2.15.2), are often inserted in private contracts. The point is, as a contractor, your concern should be with building a project. Your attorney's concern should be that the proper procedures are in place to insure that you are paid for the building of that project.

§ 2.14.8 Waiver by Agent

If your contract provides for written authorization for extra work, who may waive that provision? Certainly, the owner. He or she provided the contract, he or she almost certainly inserted the provision for written authorizations. If anyone can waive the provision, it is the owner. The owner's agent can waive the provision. Care must be exercised here. The owner's agent must have full authorization from the owner to make changes to the contract.

Can the architect or the engineer waive provisions for written authorization? This is an especially important question for you, since the architect or the engineer is usually your only contact on the project. Case law has generally held that the architect or engineer, without explicit authorization, cannot waive the contract provisions. The cases are both illustrative and instructive.

If a contract provision requires a written authorization for extra work, you must comply with that provision. The courts will not allow you to recover from the owner even if the extra work was performed pursuant to an oral directive of the architect or engineer.

When is there waiver by the owner? In *Minkus v. Sarge*, 348 Mich. 415, 83 N.W.2d 310 (Michigan), Sarge was the owner of an industrial building and had contracted with Minkus for renovation of that building. The contract required that alterations to the plans and specifications are to be by order of the owner in writing. Minkus performed various extra work including removal of a septic tank and the installation of a new roof on a portion of the building. Although no written change orders were issued, Sarge was regularly on the premises during the time the extra work was performed. The court found a waiver of the writing requirement. It is difficult for an owner to deny payment for extra work performed in his or her presence and for his or her benefit.

To secure payment on a waiver theory, what do you need to show? For starters, you should show that the extra work was actually performed. *Holdsworth v. Charles T. Wills, Inc.*, 254 A.D. 677, 3 N.Y.S.2d 226 (2nd Dept. 1938), aff'd 279 N.Y. 768 (1939), concerned a subcontract in connection with the construction of certain buildings at the Veterans' Administration facilities at Togus, Maine. Holdsworth was the subcontractor for metal furring, lathing, and plastering in the buildings. Holdsworth sought to recover from Wills, the general contractor, the costs of extra material and labor, which Holdsworth furnished in plastering the surface of the interior walls. The work was caused by an alleged defect in the construction of the walls. Holdsworth's claim for extra was stricken on two grounds. The first was that its contract required written authorization for extras and none had been obtained. The second, and more compelling reason, was that Holdsworth had not established that the extra work was actually performed.

In a similar spirit, the courts will not allow a contractor to be reimbursed for extra work, even when authorized by the owner, if the contractor causesd the need for the extra work. In *A.R.C. Industries, Inc. v. State*, 551 P.2d 951 (Alaska), A.R.C. was to build a weir across a stream bed. Had A.R.C. acted with due diligence, the project could have been completed in the summer months. A.R.C. did not, and its delay extended the contract work into the rainy season. The weir site was scoured during a storm. The State gave authorization for, and the contractor installed, an additional 1,335 cubic yards of rock to provide the proper foundation for the weir. The court refused to allow a payment on this extra. A contractor should have been aware of the effect heavy rains would have on the contract site and should have taken this into consideration in its scheduling and progressing of the work.

The general principles defining the responsibility of an agent will define the scope of his or her authority on a construction project. Where an agent is authorized by power of attorney to act for the owner in executing a building contract, the owner will be held liable for his or her acts in all matters regarding the building construction, including contractual waiver. *Michaud v. MacGregor*, 61 Minn. 198, 63 N.W. 479 (Minnesota).

It has long been the law in Nebraska that notwithstanding a contractual provision requiring written authorization of extra work, recovery will be allowed where there is evidence of waiver. In *McLeod v. Genius*, 31 Neb. 1, 47 N.W. 473 (Nebraska 1890), recovery for extras was allowed upon the mere oral request of the owner's agent. The agent agreed that the contractor would be paid for the work and the agent realized that the work would not be done without extra compensation. The agent assented to the work and permitted the work without objection. The court concluded that payment to the contractor was justifiable on two grounds. Initially, the court concluded that there was a waiver of the contract provision. In addition, the court felt that an executed oral agreement had occurred for extra work beyond the limits of the original agreement. The law in Pennsylvania is the same. See *Universal Builders, Inc. v. Moon Motor Lodge, Inc.*, 430 Pa. 550, 244 A.2d 10 (Pennsylvania).

What course of conduct will indicate waiver? The contract in *Walsh v. Howard & Childs*, 61 Misc. 328, 113 N.Y.S. 499 (App.T. 1908), required that any authorization for extra work would be signed by the owner or the owner's attorney, duly authorized by sealed power of attorney. Walsh, who was constructing and repairing a boiler house, offered four examples of waiver. In the first instance, Walsh had submitted a proposal and estimate on the letterhead of the owner, which was signed by the owner's manager. The work was done, and it was paid for. In the second instance, a written estimate was provided on Walsh's letterhead and signed off by the owner's manager. As the court points out, there was no proof that the owner's manager was duly authorized by power of attorney. In the third instance, extra work was performed relating to the foundation of the boiler house. There was no written acceptance, although the work was accepted in writing after the fact and paid for.

With respect to these three instances, the court, rather than finding waiver, felt that the facts indicated a consistent observation of the contractual provision requiring written authorization.

The fourth instance was at issue before the court. A written proposal and estimate was prepared by Walsh but was never accepted in writing by the owner, by his manager, or by anyone on his behalf. The court, strictly interpreting the contract, held that even assuming that the prior three instances showed a waiver for each particular case, those instances would not apply to the case before the court. The court deemed each of the prior three instances to be independent transactions. Howard & Childs was free to insist on strict compliance with the contract provisions in the fourth instance.

Although this is a harsh result, it appears that the court was not fully convinced that the prior three instances would indicate waiver of the contractual requirement. If the prior three instances had been deemed by the court to indicate wavier, then perhaps the court would have considered a course of conduct to have been established and found for the contractor.

Arrow Plumbing Co. v. Dare Constr. Corp., 212 N.Y.S.2d 438 (Nassau Co. 1961), involved a subcontract for plumbing work in a new home. The subcontract stated that no additional work was to be done unless authorized in writing. The court found that this clause was inserted for the protection of the plumbing subcontractor to excuse his nonperformance of any work in the absence of a written order. Since the clause was inserted for his protection, he is free to waive that clause. Consequently, the plumbing subcontractor was allowed to collect for these extras even though there was no written authorization.

In *Traitel Marble Co. v. Brown Bros., Inc.*, 159 A.D. 485, 144 N.Y.S. 562 (1st Dept. 1913), a dispute arose over a change in a building contract from hydraulic tile to quarry tile. Here, the contract required that no extras were to be allowed except upon written order of the architect. The curious twist in this case is that the architect was also the secretary of Brown Bros., Inc., the corporate owner. Brown, as secretary of the corporation, had presumed authority as broad as that of the president of the corporation. However, the court found that Traitel Marble was put on notice, pursuant to the written agreement nominating Brown as architect for the work in question, that his general authority to act as an officer of the corporation was withdrawn and limited expressly to change orders authorized by him in writing. Thus, when Brown authorized substitution of quarry tile, he was representing Brown Bros. as an architect and not as secretary of the corporation. He had no authority to direct the plaintiff orally. The contract provision requiring a writing prevailed. This case illustrates that the authority of an architect or an engineer must be carefully questioned by you.

Likewise, in *Massachusetts Bonding & Ins. Co. v. Lentz*, 40 Ariz. 46, 9 P.2d 408 (Arizona 1932), an architect ordered changes in the construction of a building. This was insufficient to indicate waiver of the provision that no change would be made unless pursuant to a written order from the owner, signed or cosigned by the architect or, in the alternative, an order from the architect stating that the owner had authorized the change. The court indicated that the authority of the architect was limited by the provisions in the contract. The contractor was on notice as to that limitation. See also *McNulty v. Keyser Office Bldg. Co.*, 112 Md. 638, 76 A. 1113 (Maryland, 1910); *Citizens Nat. Bank v. L.L. Glascock, Inc.*, 243 So.2d 67 (Mississippi).

A contractor was denied compensation for extra work performed where the contract required that extra work need be done pursuant to a written order from the architect, countersigned by the owner. In *Kelly v. St. Michael's Roman Catholic Church*, 148 A.D. 767, 133 N.Y.S. 328 (2nd Dept. 1912), there admittedly were no writings. Kelly claimed waiver of the provision. In this case, the court found that the provision was inserted in the contract for the benefit of the owner and that it could be waived by the owner or by his conduct. However, such waiver could only be by the owner or by his duly authorized agent. An architect is not to be construed as an agent of the owner for the purposes of waiving clauses in the contract. Although Kelly testified that the architect assured him that change orders would be sent, they never were. Additionally, Kelly testified that the rector of St. Michael's told him that he knew there was extra work, that Kelly hadn't received change orders, that the extra work was being done, and that the rector would see that Kelly was paid. The court stated that it searched the record in vain for an instance of waiver by St. Michael's. There was no such instance, and Kelly's extra work claim was denied.

Woodruff v. Rochester & P.R. Co., 108 N.Y. 39 (1888), involved the construction of a portion of a railroad. Woodruff entered into a subcontract with Thomas & Company, who had a subcontract with Brown Howard & Company, who had a contract with Rochester and Pittsburgh Railroad. Woodruff's subcontract required that extra work must be done pursuant to a written order from the engineers in charge. Brown Howard had hired and paid for all the engineers in charge.

In order to lay the bed of the railroad at a certain point, Woodruff needed to perform significant excavation. When the excavation was substantially complete, large amounts of earth fell from the sides and caved into the cut at four different times. Woodruff excavated and removed the earth each time. Plaintiff showed that the removal of the earth was done at the request of the engineers in charge of the work. However, the court pointed out that there was no evidence showing that the engineers had any authority to bind the railroad for this as extra work or to bind the railroad for the work as an independent contract. In fact, the court found insufficient proof that the railroad in any way ratified the agreement made by Woodruff with the engineers.

The court found that the engineers in charge had no power or authority to alter or change the contract provision, which required that extra work be memorialized in a writing. If engineers are deemed to be agents of the railroad, there are agents with special limited powers to do engineering work, to superintend, or to direct as to the execution of the contract. As limited agents, they would have had no power to alter or vary the terms of the contract or create obligations binding upon the railroad that were not embraced in the contract itself. The fact that the railroad accepted the work and had benefit of the work did not amount to a ratification of the engineers' ordering of the work. See also *Vanderwerker v. Vermont C.R. Co.*, 27 Vt. 130 (Vermont, 1854); *Devenow v. St. Peter*, 134 Vt. 245, 356 A.2d 502 (Vermont).

In *Citizens Nat. Bank v. L.L. Glascock, Inc.*, 243 So.2d 67 (Mississippi), an engineer, who was an employee of the owner's architect, approved extra work. The contract expressly limited the authority of the architect and also contained a provision requiring that extra work orders be signed by the owner or authorized by an architect's statement that the owner had approved of the work. The court denied the contractor's claim for extra work.

2.15 MISCELLANEOUS PROVISIONS

§ 2.15.1 No Damages for Delay

One of the most controversial provisions in a construction contract is the "no damages for delay" provision, which has been present in contracts for the past century. However, the law has evolved in the interpretation and application of that provision. Unfortunately, for you as the contractor, the evolution has been decidedly pro-owner; especially if that owner is a public agency.

Because the controversy has been most acute in New York, this discussion will follow the evolution of New York case law, which reflects the evolution nationwide. Of course, variations in other states will be noted.

In the categorization of contract provisions, no damages for delay is considered an exculpatory provision, that is, a provision that "takes away the blame." As a general rule, the law disfavors exculpatory provisions and construes them strictly against the person proposing the provision. In our case, that would be the owner.

The classic no damages for delay provision states that the contractor can make no claim for damages in connection with the delay in performance of the contract because of any act or omission by the owner. The provision continues that any claim for delay shall be fully compensated by an extension of time to complete performance of the work.

Does this provision give anything to you as a contractor? First, let us consider what it takes away. Where an owner causes a delay in the absence of a no damages for delay provision, he or she is responsible for damages that naturally flow from that delay.

What are those damages in a construction contract? Labor rates may increase over the course of the delay. The owner would be responsible for that increase. Materials costs may increase over the period of the delay. The owner would also be responsible for that increase. The owner would also be responsible for the contractor's extended home office overhead for the period of the delay and his or her extended field office overhead. If acceleration of the job occurs due to the delay, the owner would be responsible for acceleration costs. If the delay causes an efficiency loss, the owner would be responsible for that lost efficiency. All these real expenses are wiped out by a no damages for delay provision.

Returning to the original question, what does the contractor gain? The extension of time relieves the contractor from the threat of liquidated damages. Of course, since the liquidated damages provision has probably been inserted in the contract by the owner, this is no real gain to the contractor. In fact, the courts would not enforce a liquidated damages provision if it was shown that the contract was not completed on time due to the acts or omissions of the owner. The extension of time allowance in the no damages for delay provision is merely returning to the contractor what the owner has previously taken away. There is no gain, but an actual loss of the labor, material and overhead increases suffered due to an owner-induced delay.

Public agencies and their legal staffs have become greatly enamored of no damages for delay provisions. The best advice that can be given to you is that an appropriate contingency should be included in your lump sum bid to cover the monetary impact of an owner-induced delay.

Without a no damages for delay provision, the owner in the construction contract would be liable for damages on the legal principle that the owner has an implied obligation not to prevent the performance of the contract by unreasonable delay. In the absence of a no damages for delay provision, the owner would be liable for any delay that he or she caused.

The case law as it had existed in New York for decades was that a contractor could recover despite the broad wording of a no damages for delay provision if the delay was not within the contemplation of the parties at the time of contracting or if the delay was occasioned by conduct of the owner, which constituted active interference with the contractor's performance.

This proposition is exemplified by *Peckham Road Co. v. State*, 32 A.D.2d 139, 300 N.Y.S.2d 174 (3rd Dept. 1969), aff'd 28 N.Y.2d 734, 321 N.Y.S.2d 117 (1971). Peckham Road had entered into a contract to construct a new highway and to reconstruct portions of an existing state highway. The new highway grade proceeded through two properties, both occupied by their owners at the time of the execution of the contract. Peckham Road sued the state for the delay in obtaining possession of the property. Peckham Road had started work in March. One of the properties continued to be occupied until September. The state relied on a special note that indicated possession of buildings within the highway limits was not immediately available. The provision also stated that negotiations for the premises were underway and that the premises would be made available to the contractor as soon as they were vacated. The trial court awarded Peckham Road almost $200,000 for increased costs incurred due to the delay.

The trial court noted the well-established rule that a contractor is entitled to a reasonable opportunity to perform his or her contract without obstruction or interference. That court further stated that the owner would be held liable for failure to furnish such opportunity. Delay damages are actionable if not within the contemplation of the parties at the time the contract is made. The appellant court noted, however, that the alleged delay in *Peckham Road* was precisely provided for in the contract. That court found that the delay against which the state sought to protect itself was in the contemplation of the parties at the time the contract was signed.

The law had always been that a no damages for delay provision would not bar claims for delays if the delays were not within the contemplation of the parties. In *Peckham Road,* the appellate court found that the contract did expressly address the issue of delays in acquisition of the properties. The highest court of the state affirmed.

The *Peckham Road* criteria is simple enough. An exculpatory no damages for delay provision will be enforced where the delay is within the contemplation of the parties and there is no active interference. How is the contemplation of the parties determined? By reviewing the contract document. In *Peckham Road*, it was specifically stated that certain properties had not been acquired and that those properties would be turned over to the contractor once they were acquired.

The criterion is fair. The owner who drafted the contract and who is attempting to excuse himself from delays must itemize what those delays are to be. The contractor, seeing those delays, is in a position then to include the appropriate contingency in his bid. Public owners were not happy with the results of *Peckham Road*. In many cases, delaying factors were not specified and the contractor collected delay damages.

The tide began to turn against municipal contractors in *Kalisch-Jarcho, Inc. v. City*, 58 N.Y.2d 377, 461 N.Y.S.2d 746 (1983). Kalisch-Jarcho was a successful bidder for a heating, ventilating, and air conditioning contract in connection with the construction of a new New York City police headquarters. Its contract, for slightly in excess of $8,000,000, was fixed at 1,000 consecutive calendar days. The contract was extended for an additional 20 months. At trial, Kalisch-Jarcho presented evidence as to the City's endless revisions of plans and drawings, its failure to coordinate the activities of other contractors on the site, and acts or omissions interfering with the sequence and timing of Kalisch's work. The jury awarded Kalisch delay damages in excess of $800,000. An appellate court affirmed and the City of New York appealed to the State's highest court. The court noted that a "no damages" provision will not exonerate a party from liable under all circumstances. It will not, for example, apply to an owner acting in bad faith or to willful, intentional, or grossly negligent acts.

At trial, the jury determined that the delays were contemplated at the time of contracting, but that the delay caused by the City constituted active interference. The State's highest court left intact the principle that if the delay is not within the contemplation of the parties, then a no damages provision is unenforceable. However, the court found that once the delay was within the contemplation of the parties, then the no damages provision could not be circumvented by claiming that the delay was caused by the City's active interference. Active interference is no longer enough. In order for the provision to become unenforceable, the contractor must prove bad faith evidenced by willful, intentional, or grossly negligent conduct. In the past, the no damages provision would be enforced if the delay was within the contemplation of the parties, unless that delay was caused by active interference by the owner, with the contractor's performance. *Kalisch-Jarcho* changes the settled meaning of this contractual term when it finds that active interference is not enough. The interference must also evidence bad faith, or willful, intentional, or negligent conduct.

What was the result of *Kalisch-Jarcho*? Those contractors, engaged in public enterprises, had a need to know, with reasonable certainty, the legal consequences of their actions. The law was now in a state of flux. The agencies and the public contractors, who were previously aware of the consequences of this standard contractual term, repeatedly interpreted and applied by the courts, were left standing on shifting sands. The dissent in *Kalisch-Jarcho* suggested that if the City was not willing to accept the previous interpretations of the contract terms, it should have altered the contract.

While a no damages for delay provision will be strictly construed, the courts will not hesitate to enforce the provision. Although the application may seem harsh, both parties had freely entered into the contract and it will be enforced by the courts. See *Cunningham Bros., Inc. v. Waterloo*, 254 Iowa 659, 117 N.W.2d 46 (1962).

In *Southern Gulf Utilities, Inc. v. Boca Ciega Sanitary District*, 238 So.2d 458 (Florida App. 1970), the court held that a no damages provision could protect the District against ordinary delay in securing rights of way. However, the willful failure of the District to provide a right of way will not allow it to hide behind a no damage clause. See also *Ace Stone, Inc. v. Wayne*, 47 N.J. 431, 221 A.2d 515 (New Jersey 1966).

The most important progeny of *Kalisch-Jarcho* was *Corinno Civetta Constr. Corp. v. City of New York*, 67 N.Y.2d 297, 502 N.Y.S.2d 681 (1986), on remand 123 A.D.2d 571, 507 N.Y.S.2d 146 (1st Dept. 1986). Pursuant to its contract with the City, Civetta was to reconstruct and install approximately 245 feet of combined sewer beneath the west side of Central Park. The contract, for approximately $200,000, was executed on September 13, 1979 and was to be completed within 100 consecutive days. The work was not completed until December 1980. Civetta sought damages for $162,000. Just prior to the execution of the Civetta contract, the City had declared a moratorium on street openings. Civetta's damages were the costs resulting from the City's refusal to authorize Civetta to make additional expenditures for overtime work to complete the entire project before the moratorium took effect. Because overtime was not authorized, Civetta could not complete the project until after the moratorium was lifted. The contract, however, specifically stated that no work would be permitted during the moratorium. The court, therefore, found that this delay was contemplated by the parties. The court found no issue of gross negligence, bad faith, or intentional misconduct. The City's contention in *Civetta* was that its no damages for delay provision would bar any claim for delay, whether contemplated or not. The court rejected this argument and itemized four specific instances in which damages could be recovered: (1) if delays are caused by the owner's bad faith, evidenced by willful, intentional, or grossly negligent conduct; (2) if the delays are uncontemplated; (3) if delays are so unreasonable as to constitute an intentional abandonment of the contract; and (4) if the owner's delays amount to a breach in the fundamental obligations of the contract. A delay causing a breach of fundamental obligations is not the ordinary "garden variety" delay.

In *Buckley v. City*, 121 A.D.2d 933, 505 N.Y.S.2d 140 (1st Dept. 1986), appeal dismissed 69 N.Y.2d 742, 512 N.Y.S.2d 1030 (1987), the completion of a pumping station was delayed for eight years due to the City's improper design of a cofferdam. The court held that the no damages provision barred Buckley's claim for delay damages. The court found that the City's negligence was ordinary, not gross. This was a garden variety delay. In addition, the court found that the delay was contemplated because the contract provided a procedure to modify the agreement for additional work caused by unanticipated subsurface conditions. The improperly designed cofferdam was, for this argument, an unanticipated subsurface condition.

The court made the determination that, as a matter of law, mere reference to an unanticipated subsurface condition means that the improper design was to be contemplated. It finds the length of the delay to be irrelevant.

In *Buckley*, the line of reasoning initiated in *Kalisch-Jarcho* is brought to its absurd conclusion. It is difficult to understand how either of the parties could have contemplated an eight-year delay. Yet, in *Buckley*, the court states that as a matter of law, the delay was contemplated.

The law has come a long way since ,*American Pipe & Constr. Co. v. Westchester County*, 292 F. 941 (2nd Cir. 1923). There a public agency was liable to the contractor for a delay of 258 days in providing a right of way. The court said that any time over three months was unreasonable.

A no damages for delay provision was held to be not against public policy in *A. Kaplen & Son, Ltd. v. House Authority of Passaic*, 42 N.J. Super. 230, 126 A.2d 13 (New Jersey 1956). Nor does a no damage provision indicate a voidable adhesion contract. *W.C. James, Inc. v. Phillips Petroleum Co.*, 485 F.2d 22 (Cir. 10th Colorado 1973).

An inventive contractor attempted to circumvent a no damages provision by asserting that it applied only to delays during the progress of the work, but not to delays occurring before the work began. The argument was rejected in *Nelson v. Eau Claire*, 175 Wis. 387, 185 N.W. 168 (Wisconsin 1921). The contractor was not entitled to recover for delay by the city in removing streetcar tracks and gas mains from an old bridge obstructing the contractor's progress.

The contractor fared better in the Texas case *Dallas v. Shortall*, 87 S.W.2d 844, rev'd on other grounds 131 Tex. 368, 114 S.W.2d 536 (Texas Civ. App. 1935). The commencement of a tunnel construction project was delayed by failure of the city to acquire a right of way. The court held that such failure was not contemplated. The court stated that at the time the city directed construction to begin, it had no right to construct a tunnel. Here, the contractor had a right to, and did, rely on the city's fundamental duty to secure a right of way before the work could begin. The contractor prevailed in spite of the no damages provision.

In *Coleman Bros. Corp. v. Commonwealth*, 307 Mass. 205, 29 N.E.2d 832 (Massachusetts1940), delay damages were not recoverable in connection with a bridge construction contract. The delay was caused by time lost in the owner's considering the desirability of a foundation redesign, in the award of the draw span contract, and in the conducting of pile tests. The contract had provided that there would be no damages for delay caused by the Bridge Commission in performing or furnishing any work or materials.

In passing, it must be noted that under certain provisions, extensions of time to complete are not automatically granted under the contract terms. Where extensions of time must be requested by the contractor in writing, the absence of a writing will preclude the compensation of that contractor even though the delay was due to the owner's acts.*Quin Blare Enterprises, Inc. v. Julien Constr. Co.*, 597 P.2d 945 (Wyoming). See also *State Surety Co. v. Lamb Constr. Co.*, 625 P.2d 184 (Wyoming).

A contractor was denied recovery in *State Highway Admin. v. Greiner Engineering Sciences, Inc.*, 83 Md. App. 621, 577 A.2d 363, cert. denied 321 Md. 163, 582

A.2d 499 (Maryland 1990). The courts stated that the "uncontemplated by the parties" exception to the no damages for delay provision is not recognized in Maryland. Where there is no intentional wrongdoing or gross negligence and where the enforcement of the provision is not unconscionable, the court will enforce it. See also *Siefford v. Housing Auth. of City of Humboldt*, 192 Neb. 643, 223 N.W.2d 816 (Nebraska).

The pendulum is beginning to swing in the other direction, however. An Illinois court noted various exceptions to the "no damages for delay" provision. If the owner is not acting on good faith; if the delay is unreasonable in duration; if it is not within the contemplation of the parties; if it is attributable to inexcusable ignorance or incompetence of the engineer, then such a provision is inapplicable. *Bates & Rogers Constr. Corp. v. North Shore Sanitary Dist.*, 92 Ill. App.3d 90, 414 N.E.2d 1274 (Illinois).

There is a certain deference afforded a public agency. In *Earthbank Co., Inc. v. City*, 568 N.Y.S.2d 101 (1st Dept. 1991), Earthbank had contracted with the City of New York to construct a football and soccer field. The contract required that Earthbank excavate the entire site, level the bottom area, and refill it. Part of the site contained a wetlands area. A permit from the state was required before excavation could commence. Earthbank had commenced excavation without a permit and was ordered to cease. The contract contained a no damages for delay provision as well as a provision obligating Earthbank to obtain "all required permits." Earthbank argued that the City of New York was obligated to obtain the wetlands permit. Ultimately, the City did obtain it. Earthbank contended that the duration of the contract was insufficient if Earthbank was to have obtained the wetlands permit. The court rejected this argument. The City's failure to obtain a wetlands permit did not constitute an implied breach of a fundamental obligation. The court noted that delay damages may be recovered even in the presence of a no damages provision, where there is a breach of a fundamental obligation. However, the court stated that the fundamental obligation must be expressed, not implied. Earthbank could not recover for its delay damages.

In *J.J. Flannery, Inc. v. Kerby Saunders, Inc.*, 570 N.Y.2d 44 (1st Dept. 1991), Flannery, a sheet metal subcontractor, sued Kerby Saunders, the heating, ventilating, and air conditioning contractor. Flannery's subcontract contained a no damage for delay provision. The court held that a 25-month delay raised various issues of fact, including whether the no damage for delay provision was intended to immunize Kerby from its own acts causing delays. *Civetta*, which came through this very court, is cited on the issue of the length of the delay. However, Civetta never enumerated unreasonable delay as an exception to the no damages provision. The court's reasoning was, if the delay had been so great or so unreasonable that it might be deemed an abandonment of the contract, then the no damages cannot apply.

It should be noted that the parties in *Flannery* did not include a governmental agency. The ever vigilant municipal departments of law, recognizing a boon to be exploited in connection with the requirement that delays be contemplated at the time of contracting, immediately set to work. With great fanfare, a corporation counsel of one of the larger cities proposed new amendments to the no damages for delay provision. What were the causes of delay, which the contractor was to acknowledge and contemplate? The contractor was to acknowledge that delays are common on public work projects. In addition, he was to contemplate that the plans, drawings, and specifications may be incomplete or contain errors or discrepancies, which would necessitate the issuance of numerous change orders. He was to contemplate that those change orders would require extensive and time-consuming reviews by various individuals and bodies. He was to contemplate that the city would not have sufficient personnel assigned to the construction project to facilitate its efficient operation. The contractor was to acknowledge that he assumes all risk for any loss or damage due to delay, including those risks arising from an act or omission of the city or its representatives. The prior acknowledgments would have covered incompetence on behalf of the city. Did the acknowledgments cover corruption as well?

The contractor was to agree that the following specific examples were agreed to by him: the failure of the city to make a project site available or to provide access to the project site, the failure by the city to secure a timely permissions, approvals and consents with respect to utilities affected by the work, errors of design and the corrections of those errors, errors or inaccuracies in the plans, specifications and drawings, design changes, and defective and incorrect schedules of work. In addition, the contractor was to acknowledge that he had contemplated delay due to the fact that the city had entered into other contracts for electrical, plumbing, heating, ventilating, air conditioning, and general construction work. Delay was to be contemplated due to the action or inaction of other contractors due to the failure of the city to furnish efficient management and supervision of the project; due to delays caused by a consultant retained by the city; and due to delays in the city processing and approving shop drawings, samples, and other submittals. In addition, the contractor agreed that he had contemplated delays in the city's issuing of change orders, in its failure to provide a notice to proceed within the period specified in the agreement; and delays due to site or soil conditions differing from those in the contract, including unstable soil, obstructions, water from underground streams, and other sources. The contractor contemplated suspension of the work, untimely inspections, and tests and untimely issuances of approvals. He was also to contemplate delays in granting of substantial or partial completion and final completion, delays in issuing punch lists; delays occasioned by any change in law, rule, or regulation; and delays concomitant with the untimely delivery of equipment or material by the city, as well as untimely decisions, clarifications or replies by the city. Finally, as a coverall cause, the contractor was to acknowledge that he had contemplated "any other situations" even if they were not delineated in the prior categories. These were or could have been anticipated because of the situations referred to in the contract and arose out of the nature of the work. The delays were to be anticipated by the contractor because he had experienced such situations on prior projects or these situations were discussed as possibilities between representatives of the parties prior to bid.

In order to tie up nicely the other exceptions to the no damages provision, the contractor was to further agree that none of the delays were to be considered a breach of a fundamental obligation of the contract or an abandonment of the contract by the city.

This proposed amendment resulted in an uproar in the construction industry. Contractors, however, were powerless to affect the outcome. It was clear that the city had a new source of funds for its construction costs. No longer need those costs be financed by increased taxes, a clearly unpolitic suggestion. Now the increased costs on construction projects could be borne by the contractors, notwithstanding the fact that the city had caused the delays.

How was the issue finally resolved? The sureties, who issued bonds for the city's contracts, reviewed the proposed amendment and simply refused to issue any future bonds. All city contracts required bonding. Without bonding, the entire construction program of this large municipality would grind to a halt. The amendment was quietly withdrawn.

In conclusion, what advise can I give to you regarding the no damages for delay provision? Such provisions appear in many contracts. If you are working on a public contract, it will almost certainly contain a no damages provision. Speak with your attorney. The courts are more ready to enforce a no damages provision against a contractor in a public contract as opposed to a private contract. Case law remains unsettled, but a review of the provision in your contract by your attorney would put you in the best position to recognize any of the exclusions: abandonment, bad faith, breach of a fundamental obligation, or uncontemplated delays. With the proliferation of no damages provisions, the contractor's best defense is a hefty bid contingency. Unfortunate though it may be, the conduct of various public agencies forces the contractor to set aside as a direct cost to the job, not only the actual construction cost of a public project, but the cost of public project litigation.

§ 2.15.2 Payment Contingent

Yours is a construction company, not a lending institution. You were hired to construct a project, not to finance it. The question of when you are to be paid is paramount. You have salaries to meet, materialmen to pay, and overhead costs to satisfy. The payment provision of a contract should be reviewed before the decision is made to bid on the project.

Significant contentions regarding payment generally do not arise in the owner/contractor situation. Payment provisions in such a context are generally clear and although disputes may arise as to entitlement and timing, they do not reach legally cognizable levels. The major contention arises in the contractor/subcontractor payment provisions. Why is this? When you perform work as a subcontractor, you should review the payment provision carefully. Often, the subcontract will provide that payment to the subcontractor is contingent upon payment by the owner to the contractor for the work under the subcontract. That should not disturb you greatly.

As we will see below, the courts have held that your wait for the owner to pay the contractor should only be a wait for a reasonable time.

For some contractors, that has not been enough. Some contractors place in their contracts a payment provision that payment by the owner to the contractor is a condition precedent to payment by the contractor to his subcontractors. Conditions precedent are old friends. We last met them in § 2.6.2. By making payment from the owner a condition precedent to payment to the subcontractor, the contractor accomplishes one important object. The risk of nonpayment by the owner is shifted from the contractor to the subcontractor. If the owner does not pay, the contractor is not legally bound to pay the subcontractor.

When, acting as a subcontractor, you identify a condition precedent in your payment clause, it should be immediately stricken. A subcontractor should not be expected to accept the risk of nonpayment by the owner. The general contract is for a greater sum than the subcontract. The risk of nonpayment should remain with the parties to the general contract; i.e., the owner and the general contractor.

A "pay when paid" provision typically provides that the subcontractor will be paid within a stated number of days after the contractor has received payment from the owner and that the subcontractor's payments are limited to those funds actually received by the contractor from the owner for payment of work performed by the subcontractor.

The seminal case in this area is *Schuler-Haas Electric Corp. v. Aetna Casualty & Surety Co.*, 49 A.D.2d 60, 371 N.Y.S.2d 207 (4th Dept. 1975), aff'd 40 N.Y.2d 883, 389 N.Y.S.2d 348 (1976). Schuler-Haas performed electric work on two separate subcontracts. The Aetna Casualty & Surety Company provided the payment bonds for the third ward renewal project in Rochester, New York. Aetna acknowledged Schuler-Haas' status as a subcontractor. Schuler-Haas had satisfactorily performed on its subcontracts. However, Aetna refused payment, stating that pursuant to the terms of the subcontract, the general contractors were to pay Schuler-Haas when full payment for the subcontract work was received from the owner. In addition, the general contractors were to pay Schuler-Haas within 15 days after they received final payment from the owner. Aetna argued that since payment had never been received from the owner, it could not be held responsible under the terms of its payment bond. The surety could assume the same defenses which its principals, the general contractors, had. The court noted that, as a practical matter, suppliers and small subcontractors on large construction projects require reasonable prompt payment in order for them to remain solvent and stay in business. The court felt that there was no clear expression by the general contractors that delay in payment was to be shifted to the subcontractor. Consequently, the subcontracts should not be interpreted as intending such a shift. See *Thos. J. Dyer Co. v. Bishop Intern. Engineering Co.*, 303 F.2d 655 (6th Cir. 1962).

It should not be presumed that the subcontractor intended that his payments should await the determination of an extended legal dispute between general contractor and owner. *Eastern Heavy Constructors v. Fox*, 231 Md. 15, 188 A.2d 286 (Maryland 1963). Certain courts have held that a provision requiring the contractor to pay his or her subcontractor within a stated number of days after payment from the owner merely fixes the date payment is due and does not establish a condition precedent to payment. *Howard-Green Elec. Co. v. Chaney & James Const. Co.*, 12 N.C. App. 63, 182 S.E.2d 601 (North Carolina 1971); *A.J. Wolfe Co. v. Baltimore Contractors, Inc.*, 355 Mass. 361, 244 N.E.2d 717 (Massachusetts 1969); *Fishman Const. Co. v. Hanson*, 238 Md. 418, 209 A.2d 605 (Maryland 1965); *Moore v. Continental Cas. Co.*, 366 F.Supp. 954 (W.D. Oklahoma 1973).

Where the subcontract does not specify a time of payment, the courts have held that payment is due to plaintiff within a reasonable time after completion of the work and demand for payment. *Byler v. Great Am. Ins. Co.*, 394 F.2d 273 (10th Cir.).

The court in *Schuler-Haas* determined that payment from the owner was not a condition precedent pursuant to the terms of this pay when paid clause. The subcontractor collected from Aetna.

In *Action Interiors, Inc. v. Component Assembly Systems, Inc.*, 144 A.D.2d 606, 535 N.Y.S.2d 55 (2nd Dept. 1988), Action Interiors performed taping and spackling work on the interior drywall of a hotel construction project. A dispute arose among the owner, the general contractor, and Component Assembly Systems, Inc., the subcontractor for interior drywall installation. Consequently, Component Assembly Systems had not received full payment for its work from the owner. Component resisted payment by relying on the subcontract provision that payment to Action is not due until the owner has paid Component. The court held that absent a clear expression to the contrary, a pay when paid provision does not establish a condition precedent. The provision provides a postponement of payment to permit a general contractor the opportunity to obtain funds from the owner. Component Assembly Systems could not resist payment for more than a reasonable time. The court found for Action Interiors.

The pay when paid provision in *Grossman Steel and Aluminum Corp. v. Samson Window Corp.*, 78 A.D.2d 871, 433 N.Y.S.2d 31 (2nd Dept. 1980), aff'd 54 N.Y.2d 653, 442 N.Y.S.2d 769 (1981), is instructive. It provided: "Retained percentages will be paid to [Grossman] as and when [Samson] receive[s] such payment from the [State University Construction Fund] and in the same proportions." The court found that Grossman, the subcontractor, was entitled to the balance due on the subcontract with Samson despite the fact that the State University Construction Fund had not yet paid Samson. Grossman needed only wait a reasonable time for payment. That reasonable time had expired.

The result is completely different, however, where the payment clause establishes a condition precedent. A condition precedent provision will limit the general contractor's financial risk and pass that risk to the subcontractor.

A payment provision as a condition precedent was found in *Gilbane Building Co. v. Brisk Waterproofing Co., Inc.*, 585 A.2d 248 (Maryland 1991). Gilbane contracted with the owner to act as construction manager and general contractor for a project in which privately owned warehouses would be converted to residential condominiums. Gilbane subcontracted with Brisk for certain masonry work. The payment provision stated, "it is specifically understood and agreed that the payment to the trade contractor [Brisk] is dependent, as a condition precedent, upon the construction manager [Gilbane] receiving contract payments, including retainer from the owner." Brisk completed its work and requested final payment. Gilbane refused, since it had not received payment from the owner. After the owner filed for bankruptcy, Brisk sued Gilbane. The trial court ruled in favor of Brisk. The trial court stated that under a pay when paid clause, the insolvency of an owner would not defeat the subcontractor's claim against the general contractor.

Gilbane appealed the decision of the trial court. Brisk argued that it did not intend to assume the risk of bankruptcy by the owner. The appeals court held that the condition precedent provision was clear and unambiguous. The language in the subcontract shifted the risk of the owner's nonpayment from Gilbane to Brisk. The decision of trial court was reversed.

In *Gilbane*, the owner went into bankruptcy. In *Richard De Castro & Sons Electrical Corp. v. Licon-En-Corp Joint Venture*, N.Y.L.J., Feb. 11, 1991, (N.Y. Co. 1991), De Castro was the electrical subcontractor. The Joint Venture had obtained a judgment against the owner, which remained unpaid. It asserted the payment provision that provided: "Payment of the approved portion of the Subcontractor's [De Castro's] monthly estimate shall be conditioned upon receipt by the Contractor [Joint Venture] of his payment from the Owner." The court found the Joint Venture's argument without merit. The payment provision did not establish a condition precedent to payment, but merely fixed a reasonable time for payment. The Joint Venture was obligated to pay De Castro.

You need not suffer others' follies in order to gain by their experiences. In *Atlas Door Corp. v. Elite Associates, Inc.*, N.Y.L.J., Oct. 21, 1988 (N.Y. Co. 1988), Elite Associates, Inc. was a general contractor who found it expedient to shift the financial risk of an owner's nonpayment to his subcontractor. Elite had a prime contract with the City of New York, by which the City would withhold five percent retainage of monies owed to Elite. Elite in turn held five percent on the Atlas subcontract. Elite's contract with the City was ultimately terminated. Atlas sought payment for its remaining five percent. Elite resisted, claiming that Atlas could not be paid until Elite received the retainage the City was holding. The court rejected this argument and decided that Elite was to pay Atlas; but Elite had learned its lesson.

In *Crown Plastering Corp. v. Elite Associates, Inc.*, 166 A.D.2d 495, 560 N.Y.S.2d 694 (2nd Dept. 1990), Elite was acting as general contractor for the Nassau County Department of Public Works in connection with the construction of the Nassau County District Courthouse. Crown Plastering Corp. had entered into a written subcontract to provide lathing and plastering work on the Courthouse. Again, Elite was terminated. Crown sought to recover the balance due on its subcontract and the value of extra work. Elite again resisted payment, arguing that payment would not be due to Crown until Elite had received its payment from Nassau County.

However, the payment provision of the subcontract now read: "receipt of payment from the Owner for the subcontractor's work is a condition precedent to payment by the Contractor to the Subcontractor. The Subcontractor hereby acknowledges that it relies on the credit of the Owner, not the Contractor, for payment of its work." The court noted that the law was well settled. Absent a clear expression in the contract that a credit risk is meant to be shifted to the subcontractor, the meaning of a provision that payment is not due until the owner has paid the general contractor merely fixes a time for payment. However, the court found that the payment provision clearly manifested the parties' intention to shift the credit risk to the subcontractor. Generally, Elite could not rely on the failure of Crown to perform a condition precedent where Elite had frustrated or prevented the occurrence of the condition precedent; e.g., preventing final payment by the city by Elite's actions resulting in a termination. However, Crown had not offered any evidence to establish that the prime contract was terminated due to Elite's breach of contract. Without such a showing, the condition precedent prevents Crown from collecting until Elite is paid by Nassau County.

Before closing, we should note one final variation in the payment provision. Often contracts will contain a "pay when approved" provision. This provision makes final payment to the contractor contingent upon approval of the work by the owner or his or her agents. This pay when approved provision does not create a condition precedent to payment where the subcontractor has performed his work and is in no way responsible for the owner's failure to approve.

The important distinction in a payment provision is this: the presence of the words "condition precedent" should immediately alert you, when acting as a subcontractor, that the risk of nonpayment by the owner is being shifted to you. That risk initially rests with the general contractor and is based upon his or her contract with the owner. The courts will require explicit language in the subcontract to recognize a shifting of that financial risk and consequently will rebuff most efforts. However, if the term of art condition precedent is used, the courts will deem the financial risk of nonpayment to have been shifted to the subcontractor. If you have any doubt as to the language contained in the subcontract, you should immediately contact your attorney.

§ 2.15.3 Arbitration

If it is arbitration that you desire, then an arbitration provision should be included in any contract. It is not an absolute necessity, of course. The parties can agree to arbitration at any time. Arbitration provisions in construction contracts typically refer to arbitration by the American Arbitration Association pursuant to the Construction Industry Rules.

There are two points worth considering. Before contract signing and while still in negotiating stage, you may request more than one arbitrator. I generally prefer to have three. A number of arbitrators will dilute the influence of a sole arbitrator who may, for whatever reason, be prejudiced against you. In addition, at the time of the negotiations, and prior to signing, the place of the arbitration should be specified. Perhaps your worst experience will be to demand arbitration and then find that it is determined that the site of the arbitration will be 600 miles away. There is no need to elaborate on the temptation placed before a local arbitrator when deciding a claim between you and your attorney and a local businessperson and his or her local attorney. If you can negotiate that home-court advantage, negotiate it now, before disputes begin.

What is the duration of an arbitration provision? The arbitration provision is generally co-terminus with the duration of the contract. In *Shawnee Hosp. Auth. v. Dow Construction, Inc.*, 812 P.2d 1351 (Oklahoma 1990), the contract contained an arbitration provision. However, the parties proceeded, not in arbitration, but in litigation. During trial, a settlement was reached, which excluded subsequent claims arising from latent construction defects. After five months, the Hospital Authority sought to reopen the litigation, alleging certain latent defects. Dow moved to compel arbitration.

The court reasons that the settlement agreement is intended to supersede the contract. The agreement stated that the contract "is hereby terminated with respect to any further performance obligation..." The settlement extinguishes the contractor's contractual obligations except for those concerning subsequent latent defects. The balance of the contract, including the arbitration provision, was discharged. The court rejected the contractor's attempt to compel arbitration.

§ 2.15.4 Exculpatory Clauses

Exculpatory clauses were discussed briefly in § 2.15.1, in connection with no damages for delay provisions. Exculpatory language "takes the blame away" from the party with whom it legally rests. Such a taking away or shifting of the blame can be accomplished in two ways: by contract or by statute. Our discussion is limited to contractual exculpatory provisions.

It should be noted that the law does not favor the shifting of blame. Contractual exculpatory provisions are strictly construed against the person from whom the blame is to shift. Where the exculpatory provision seeks to exempt a person from liability for negligence, it is strictly construed against the person relying on the provision.

A party to a contract will not be excused from his or her own negligent acts unless such intention is expressed in unequivocal terms in the contract. *Glomboski v. Baltimore & Ohio Railroad*, 35 A.D.2d 772, 316 N.Y.S.2d 904 (4th Dept. 1970). When the court searches a contract for unequivocal terms, what does it look for? A general answer can be found in cases involving the installation and maintenance of burglar alarm systems.

Dubovsky & Sons, Inc. v. Honeywell, Inc., 89 A.D.2d 993, 454 N.Y.S.2d 329 (2nd Dept. 1982), was just such a case. Honeywell had agreed to install and maintain a burglar alarm system in Dubovsky's warehouse for the storage of food stuffs. A burglary subsequently occurred. The exculpatory provision contained the following language:

> ... the Contractor is not assuming responsibility for any losses which may occur even if due to Contractor's negligent performance or failure to perform any obligation under

this Agreement. THE CONTRACTOR DOES NOT MAKE ANY REPRESENTATION OR WARRANTY, INCLUDING ANY IMPLIED WARRANTY OF MERCHANTABILITY OR FITNESS THAT THE SYSTEM OR SERVICE SUPPLIED MAY NOT BE COMPROMISED OR THAT THE SYSTEM OR SERVICES WILL IN ALL CASES PROVIDE THE PROTECTION FOR WHICH IT IS INTENDED.

In finding for Honeywell and determining that the exculpatory provision was enforceable, the court noted that where the provision is intended to insulate a party from his or her own negligence, it will be enforced when expressed in clear and unequivocal and unmistakable language. Recognize, however, that when such provisions purport to extend to willful or grossly negligent acts, they are viewed as wholly void. Dubovsky complained of ordinary, "garden variety" negligence. Therefore, the exculpatory provision is valid. See also *Florence v. Merchants Central Alarm Co., Inc.*, 51 N.Y.2d 793, 433 N.Y.S.2d 91 (1980).

When a member of a yacht club fell from a gangplank, the club contended that her claims were barred by an exculpatory provision in a membership application. The courts found the language insufficient. In *Hertzog v. Harrison Island Shores, Inc.*, 21 A.D.2d 859, 251 N.Y.S.2d 164 (1st Dept. 1964), Ms. Hertzog fell from a gangplank, landed on some rocks, and was injured. Her membership application contained the statements: "... I agree to become bound by any and all laws of the club and by all rules and regulations as they now exist, or if they may be amended and waive claim for any loss to personal property, or for any personal injury while a member of said club." The language was deficient in that it did not explicitly absolve Harrison Island from its own negligence.

In *Shalman v. Bd. of Ed.*, 31 A.D.2d 338, 297 N.Y.S.2d 1000 (3rd Dept. 1969), the court reiterated the tenet that exculpatory provisions are strictly construed against the person seeking exemption from liability. The Board of Education sought to be excused from liability, relying on a primitive no damage for delay clause. The court noted that there is a duty imposed on an owner not to interfere with the prosecution of the work of his contractor. The court also stated that the owner implicitly agrees that the contractor will not be unreasonably delayed by failure of other contractors to perform work. The court notes that for a breach of that duty, a contractor may recover resulting damages. Admittedly, this case is pre-*Kalisch-Jarcho*, but it does state the general principles regarding exculpatory provisions. For post-*Kalisch-Jarcho* analyses of exculpatory no damages provisions, see *Dal Constr. Corp. v. New York*, 108 A.D.2d 892, 485 N.Y.S.2d 774 (2nd Dept. 1985); *Martin Mechanical Corp. v. P.J. Carlin Const. Co.*, 132 A.D.2d 688, 518 N.Y.S.2d 166 (2nd Dept. 1987).

§ 2.15.5 No Changes

The "no changes" provision is an attempt by the owner to limit the amount paid on a lump sum contract to the sum specified in the contract. It is exculpatory in the highest degree. It attempts to ignore the multitude of changes which generally occur in a large construction contract. It also attempts to ignore any delays which may occur. Where a contract contains a liquidated damage clause, the court will generally not enforce a "no changes" provision unless it allows for an extension of time due to delays. Where changes orders or delays are the result of acts by the owner, an exculpatory no changes provision will be strictly construed against that owner.

§ 2.15.6 Notice

In construction contracts, the most important provisions regarding notice are those with respect to notices of claim. As explained in § 2.14.7, where the notice of claim requirement is found in the law itself, the courts will hold that the filing of such notice of claim is a condition precedent to recovery.

The notice of claim provisions in private contracts are found in the contracts themselves. The courts are more likely to find a waiver of a contractual notice of claim provision, since even a contractual provision making the notice of claim a condition precedent to payment may be modified by future agreement between the parties. Can the parties by their actions indicate a waiver of a condition precedent? Without a written agreement, this would present a difficult evidentiary problem.

§ 2.15.7 Place of Performance

The place of performance of a construction contract is generally not at issue. However, in connection with the development of lands out of a state, a few general observations are in order. The general rule is that matters relating to the performance of a contract are governed by the place where the contract is to be performed. In addition, if no place of performance is mentioned in the contract, it is presumed that the parties intend it to be performed where it is made. Contracts referring to the transfer of title of land are governed by the law of the place where the land is situated. See *Thompson v. Lakewood City Development Co.*, 105 Misc. 680, 174 N.Y.S. 825 (Kings Co. 1919), aff'd 188 A.D. 996, 177 N.Y.S. 926 (2nd Dept. 1919).

§ 2.15.8 Provision forForfeiture

Provisions for forfeiture in a construction contract can take many guises. Such a provision will generally allow the owner to seize certain of the contractor's property or to reduce the monetary amount owed the contractor. These two examples are of particular note. In *Lehmann v. State*, 123 A.D.2d 604, 507 N.Y.S.2d 15 (2nd Dept. 1986), appeal denied 69 N.Y.2d 612, the State's contract allowed for seizure of certain of the contractor's property following termination of the contract. Once termination occurred, the State had seized the contractor's personal office trailer. He argued that the trailer was not on the construction site and not properly seized by the State. The court noted that the trailer was parked at a location specifically requested by the contractor in an area adjacent to the construction. In addition, a second trailer, provided under the contract as a field office for the State's on-site engineer, was even more removed from the construction site. Furthermore, there were no contract limit lines on the plans. Under the circumstances, the court found that the office trailer was on the construction site and that the seizure was permitted.

In *C.D.E. Air Conditioning, Inc. v. New York Health & Hospitals Corp.*, 105 Misc.2d 848, 433 N.Y.S.2d 549 (Civ. Ct. N.Y. Co. 1980), C.D.E. was the heating, ventilating, and air conditioning contractor for a City project. Pursuant to its contract, C.D.E. was to provide 100 person-days for minority trainees. C.D.E. did not deny that the requirement was unsatisfied. It argued that the breach was due to no fault of its own, but rather, that to the refusal and failure of the Sheet Metals Workers' Union to supply minority workers. The contract required that C.D.E. make a good faith effort to achieve the goal of 100 person-days for minority trainees. However, the contract provision continued:

> In the event that the number of training positions is less than the assigned number for any reason whatsoever, the contractor's compensation will be decreased by an amount equal to the difference between the wages and fringe benefits paid by the contractor to the trainees and the wage and fringe benefits which would have been paid to the trainees had the number and duration of the positions been as set forth.

C.D.E.'s argument of impossibility of performance may excuse its contractual breach for failing to hire the minority trainees. However, the excused nonperformance does not include an entitlement to payment for the unperformed obligation. The court found that the deduction made from the C.D.E. contract was correctly computated and properly deducted under the terms of the agreement.

§ 2.15.9 Scaffold Provisions

As I discussed in § 2.15.2, the contract can permit the transfer of the financial risk of nonpayment by the owner. The transfer of certain risks can also be accomplished through statutory means. For example, the labor law of many states shifts the burden of insuring the safety on scaffolding jobs from the workers themselves to the owners and general contractors. Section 240(1) of the New York Labor Law creates absolute liability. An injured worker need only show a violation of the labor law and demonstrate that the violation was the approximate cause of his or her injuries. The worker need not demonstrate that the precise manner in which the accident occurred or the extent of his or her injuries was foreseeable. The worker need only demonstrate that the owner's or general contractor's negligence was a substantial cause of the events that produced injury.

In *Estate of Hassett v. Trump Village Constr. Corp.*, 575 N.Y.S.2d 843 (1st Dept. 1991), Hassett was employed by Crown Plastering Corp., a subcontractor in the renovation of a commercial building owned by Trump Village and leased by Manufacturers Hanover Trust Company. Charles Construction Corp. was the general contractor. Mr. Hassett was moving materials to the roof when he fell to the pavement from a height of eight to twelve feet. Immediately after his fall, evidence showed that a plank of scaffolding was seen dislodged from the remainder of the scaffolding. There were no safety belts, nets, or other protective devices on the site. Mr. Hassett died from his injuries. The intent of the labor law was to shift the burden of insuring safety on scaffolding jobs from the workers to the owners or general contractors. Pursuant to the labor law, the court found a liability on the part of Charles Construction to Hassett. In addition, the court found that Charles Construction Corp. was liable to both Trump Village and to Manufactures Hanover Trust Company for any damages assessed again them in connection with Mr. Hassett's death.

The court based that finding on common law indemnity. Indemnity, by applying established principles of law, causes the burden, financial or otherwise, to shift from one person to another. In *Estate of Hassett*, there was nothing to suggest that either Trump Village or Manufacturers Hanover Trust had authority to supervise and control the portion of the project on which the injury occurred. Since the liability of both Trump Village and Manufacturers Hanover Trust under the labor law is established by reason of the acts or omissions of another, Charles Construction, both Trump Village and Manufacturers Hanover Trust are entitled to indemnification.

Faced with complete responsibility for the accident, Charles Construction Corp. sought indemnification from Crown Plastering Corp. Charles proceeded on two grounds: contractual indemnity and common law indemnity. The court found that Charles had no claim for contractual indemnity. A promise to indemnify regardless of the general contractor's fault is unenforceable, pursuant to the New York General Obligations Law, § 5-322.1, as a matter of public policy.

On the issue of common law indemnity, Charles would need to show that, as a matter of law, Crown was actually responsible for the accident. This Charles failed to do.

The statutory provisions regarding scaffolding generally shift the responsibility for the safety of that scaffolding and its appertinencies to one particular member of the construction team, usually the owner, the general contractor, or both. Common law indemnification may shift that responsibility back to the proper party. What Charles attempted to do in *Hassett* was to establish indemnification from Crown by contract. The provision, although it was not quoted in the case, was apparently a promise to indemnify Charles regardless of whether or not Charles' fault caused the accident. Such an indemnification is generally against public policy. It will, and did, fail.

If these scaffolding provisions are news to you, you ought to discuss them with your attorney. In addition, any indemnification provision in a contract should be reviewed by your attorney before it is signed.

2.16 JOINT AND SEVERAL OBLIGATIONS

§ 2.16.1 Liability

The question of joint or several liability arises in a situation where you and another person agree to perform a contract. That person can be a wife, a partner, or a joint venturer. The general rule, applicable in New York as well, is that you and that other person will be jointly liable.

There need not be contract language establishing a joint liability. However, there must be explicit words severing the liability, if that is your intent. An obligation by two or more persons establishes a joint duty unless the contrary is stated.

CITATIONS

A. Kaplen & Son, Ltd. v. House Authority of Passaic, 42 N.J. Super. 230, 126 A.2d 13 (New Jersey 1956) 2.15.1

A.J. Cerasaro, Inc. v. State, 97 A.D.2d 598, 468 N.Y.S.2d 204 (3rd Dept. 1983) 2.1.2

A.J. Wolfe Co. v. Baltimore Contractors, Inc., 355 Mass. 361, 244 N.E.2d 717 (Massachusetts 1969) 2.7.3, 2.15.2

A.R.C. Industries, Inc. v. State, 551 P.2d 951 (Alaska) 2.14.8

Ace Stone, Inc. v. Wayne, 47 N.J. 431, 221 A.2d 515 (New Jersey 1966) 2.15.1

Action Interiors, Inc. v. Component Assembly Systems, Inc., 144 A.D.2d 606, 535 N.Y.S.2d 55 (2nd Dept. 1988) 2.15.2

Airco Alloys Division, Airco Inc. v. Niagara Mohawk Power Corp., 76 A.D.2d 68, 430 N.Y.S.2d 179 (4th Dept. 1980) 2.1.6

Albert A. Volk Co. v. Cauldwell-Wingate Co., 272 A.D. 290, 70 N.Y.S.2d 662 (1st Dept. 1947) 2.10.1

Allen-Howe Specialties Corp. v. U.S. Constr., Inc., 611 P.2d 705 (Utah 1980) 2.12.2

Allied Chemical Corp. v. Alpha Portland Industries, Inc., 58 A.D.2d 975, 397 N.Y.S.2d 480 (4th Dept. 1977) 2.3.1

Allis-Chalmers Mfg. Co. v. Malan Constr. Corp., 30 N.Y.2d 225, 331 N.Y.S.2d 636 (1972) 2.6.2

American Pipe & Constr. Co. v. State, 247 N.Y. 150 (1928) 2.14.2

American Pipe & Constr. Co. v. Westchester County, 292 F. 941 (2nd Cir. 1923) 2.15.1

Andrew Catapano Co. v. New York, 116 Misc.2d 163, 455 N.Y.S.2d 144 (N.Y. Co. 1980) 2.3.4

Application of Seneca Falls Central School District, 117 Misc.2d 879, 459 N.Y.S.2d 689 (Seneca Co. 1983) 2.11.2

CHAPTER 3

PERFORMANCE OR BREACH

3.1 BUILDING CONTRACTS

§ 3.1.1 Approval by Owner

We turn our attention now to a provision contained in many smaller construction contracts. That provision specifies that your work must be done to the satisfaction of the owner. What does this mean?

The general rule is that such a provision is valid and enforceable. There is no recovery on the contract if honest, good faith dissatisfaction is proved.

Duffy Bros., Inc. v. Bing & Bing, Inc., 217 A.D. 10, 215 N.Y.S. 755 (1st Dept. 1926), involved a contract that Duffy had with Bing & Bing, the builder of an addition to the Hotel St. George in Brooklyn. Duffy was to perform the plastering work on the addition. The contract also provided an acknowledgment that the building was behind schedule and that should Duffy complete the work within the time specified, Bing & Bing would pay a $5,000 bonus. The court held that in order to obtain the bonus, Duffy must show that he completed the work within the time specified or that any delay was caused by Bing and not due to the fault of Duffy. Duffy must show that the work performed on the contract was satisfactory to Bing or that it should have been satisfactory to Bing. In addition, the failure to be satisfied by Bing must be shown to be unfounded, in bad faith, arbitrary, and capricious. Although Bing & Bing was to be the sole judge of the quality of work, his satisfaction with the work could not be unreasonably withheld. If Duffy complied with other requirements of the contract, he would be entitled to the bonus for work performed in a proper and workmanlike fashion, even if Bing was not satisfied with it. Where the issues involved are objective fitness, rather than satisfaction of subjective personal sensibility, an objective standard will be applied.

When is an objective standard applied? Have the plans and specifications been followed? Does the machinery operate as a reasonable owner would expect it to? Do the mechanical systems perform as they should be expected to perform? These are among the considerations that are subject to an objective standard. Where an objective standard is involved, the owner cannot withhold approval without a reasonable justification.*Haymore v. Levinson*, 8 Utah 2d 66, 328 P.2d 307.

In *Doll v. Noble*, 116 N.Y. 230 (1889), the contract required that the work be done "in the best workmanlike manner ... and to the entire satisfaction" of Noble, the owner. Doll was to accomplish polishing, staining, and rubbing of the woodwork in two houses owned by Noble. Doll was entitled to be paid for his work as long as Noble was satisfied. However, Noble could not defeat Doll's claim by arbitrarily and unreasonably stating that he was not satisfied. If the work is done according to the contract, then a reasonable man would have been satisfied, and Doll is entitled to recover; see also *Shimek v. Vogel*, 105 N.W.2d 677 (North Dakota) and *Eckert-Fair Constr. Co. v. Flabiano*, 342 S.W.2d 629 (Texas Civ. App.).

A contract provision requiring that construction work be done to the satisfaction of the owner is valid. *Enterprise Roofing & Sheet Metal Co. v. Howard Invest. Co.*, 105 Ohio App. 502, 152 N.E.2d 807.

Greenberg v. Lumb, 129 N.Y.S. 182 (App.T. 1911), involved a contract between White Iron Works and Lumb to furnish and erect a fence similar to the one presently in front of Lumb's premises. Lumb, in accepting the written proposal, wrote "same to be done in a workmanlike manner." White Iron Works was not bound to satisfy Lumb. It was only bound to do the work in a manner that ought to satisfy; see also *McKenzie v. Decker*, 94 N.Y. 650 (1884).

Marcus v. Nelson, 119 N.Y.S. 1085 (App.T. 1909), involved a contract in which Marcus was to paint certain rooms to the satisfaction of Nelson's agent. The work was not of a character where individuality of taste or fancy was involved. The court found that the work was well done. Nelson's agent could not arbitrarily or capriciously assert that he was dissatisfied.

3.2 SUFFICIENCY

§ 3.2.1 Plans and Specifications

In order to entitle you to payment, your plans and specifications must be complied with.

The specifications in *L'Hommedieu v. Winthrop*, 59 A.D. 192, 69 N.Y.S. 381 (2nd Dept. 1901), provided that the finishing coat of stucco work should be made of freshly burned shell lime and thoroughly washed coarse white marble dust. Robert Dudley Winthrop, building his McKim, Mead & White mansion on the north shore of Long Island, insisted that this thoroughly washed coarse white marble dust be as white as the Tuckahoe marble sills used in the construction. L'Hommedieu, the builder, argued that the plans and specifications originally submitted to him and upon which he made his estimate, required a plain cement exterior stucco coat, without the white marble dust finish. The court found that once L'Hommedieu had executed the contract with the specifications requiring the white marble dust finish, he was bound to perform the work in accordance with those specifications.

In *Howard Iron Works v. Pittsburgh Steel Const. Co.*, 143 A.D. 734, 128 N.Y.S. 89 (4th Dept. 1911), Howard was to furnish counterweights to counterbalance the movable span of a bascule bridge, which Pittsburgh Steel has contracted to build for the City of Buffalo. The specifications, which were a part of Pittsburgh's contract with the City of Buffalo, were made a part of Howard's subcontract with Pittsburgh. All requirements as to the counterweights to be furnished are found in these specifications. The estimated weight of the counterweight was 636,000 pounds, but that information was furnished only for the assistance of bidders. The immediately preceding paragraph of the specifications requires that "... the bidder hereunder must check the [loading requirements and general dimensions] and make his own estimate of weights and quantities, those shown being furnished only as a check upon such figures, quantities, and dimensions." The court found that if the bridge built according to the specifications required a greater counterweight for its successful operation, then Howard must furnish it. Conversely, if less than the estimated weight was sufficient, the furnishing of that amount was all Howard's subcontract required.

In *Acme Builders, Inc. v. Facilities Development Corp.*, 51 N.Y.2d 833, 433 N.Y.S.2d 749 (1980), Acme sued for the additional cost it incurred in painting portions of an existing boiler room. Acme claimed that, pursuant to its contract for construction work on a new boiler plant and alteration of an existing boiler room, it was required to paint only those areas of the boiler room disturbed by the construction.

The contract provided that the drawings will control over any contrary provision in the written specifications. Those drawings indicated that the entire boiler room should be painted. Acme Builders' claim was denied.

John T. Woodruff & Sons v. Lembeck & Betz Eagle Brewing Co., 164 N.Y.S. 594 (App.T. 1917), involved a contract in which Woodruff was to make alterations to a brewery. The plumbing specifications required that Woodruff connect the existing sewer with a 4-inch drain, continuing to the soil descent line, and branching off to the 4-inch leader as shown on plan. The plan shows that the soil pipe in the building was to be carried down below the level of the cellar floor. At that point was to be installed a trap and from there, a direct connection to the street sewer.

Woodruff argued that the existing in-house sewer was suspended several feet from the cellar floor and that all he was required to do was to make the necessary connections with that existing sewer. However, the specifications provided that the drawings and specifications were intended to be cooperative and anything omitted from the specifications as shown on the plans or vice versa, was to be done without extra charge. In addition, the specifications called for excavation for sewer pipe and for re-filling the excavation. They also required that the cellar floor be pitched toward the house trap. The court found that upon viewing the specifications and the drawings together Woodruff was required to place all sewer connections below the cellar floor.

A discussion of the scope of a subcontract is found in *Schwoerer v. Zimmerman*, 30 Misc. 800, 63 N.Y.S. 1020 (App.T. 1900). Zimmerman, a general contractor, had contracted to erect an apartment building. The plans and specifications of the general contract provided for certain galvanized iron work, including all cornices, balustrades, bay and dormer windows, and outside trim to 20 windows where shown on the plans. The contract required that a cornice be formed and that a roof furnished to store show windows. It also required that roofs be furnished to bay windows and a gutter be connected to a main cornice. Schwoerer was a subcontractor specializing in galvanized iron works. He wrote to Zimmerman and stated that he would perform the following work in accordance with the plans and specifications: all cornices, balustrades, bay windows, copper store cornice, slate roofing on mansard roof, and one skylight. Zimmerman accepted this offer. The court held that the subcontractor was not to perform all work specified in the plans and specifications, but only the work specified in Schwoerer's bid offer. Any item of galvanized iron work not included in that bid offer did not become part of the subcontract.

As long as the product of your work satisfies the specifications, you can choose your own method for achieving that product. In *Charles Meads & Co. v. New York*, 191 A.D. 365, 181 N.Y.S. 704 (2nd Dept. 1920), the court found that this principle was not only law, but common sense. The court noted that when a contractor bids, his or her estimates, which influence the bid, are necessarily based on the contractor's own methods of work. This contract involved the construction of the central library building of the Brooklyn Public Library. The city required Meads to pour the concrete foundation, slabs ,and walls in one operation as a monolith. This monolithic construction was not required by the specifications. Meads' claim against the city was upheld.

In *Thomas W. Funiucane Co. v. Bd. of Ed.*, 190 N.Y. 76 (1907), the Board sought to recover from Funiucane the savings Funiucane realized when the Board unconditionally agreed to have wood flooring kiln dried in the South. The specifications had required that all finished floors were to be thoroughly kiln dried, taken from the kiln directly to the sawing machines and then directly to the school building. At the request of Funiucane, the provision was waived by the Board. Funiucane had the flooring kiln dried in the South, where it was purchased, planed, and machined before being shipped.

The court noted that the Board had not claimed that the floor had absorbed moisture in being transferred to the school or that it was in any other way injured. Nor did the Board complain that the completed floors were inferior to flooring anticipated by the specifications. The savings were to remain with Funiucane.

Bestway Constr., Inc. v. Broome County, 107 A.D.2d 897, 483 N.Y.S.2d 843 (3rd Dept. 1985), involved a dispute with regard to the replacement of a concrete riser for a dam, which Bestway had constructed for Broome County. The specifications contained the minimum standards for concrete with respect to compressive strength, consistency, and air content of the mix. The specifications required that Bestway submit a proposed design mix for the concrete. Both 7-day and 28-day compressive strength tests were to be taken. The initial seven-day test showed a compressive strength of nearly 1,000 pounds per square inch less than the results of the seven-day test performed on the original design mix. Bestway was notified of the concern over the quality of the concrete. Bestway nevertheless continued to proceed with the pouring of the final quantities of concrete. The results of the 28-day tests represented an average of almost 700 pounds per square inch less than that required by the specifications. Broome County instructed Bestway to remove and replace the riser with concrete meeting the contract specifications. Bestway complied and claimed the increased costs for removing and replacing the riser.

The credible proof established that all testing procedures were valid and that the area of the riser failed to meet the contract requirements. Bestway was not entitled to recover the extra costs of replacing the riser.

§ 3.2.2 Effect of Defects

When you fully comply with the plans and specifications, you are entitled for the work performed. You recover your entire contract amount, even if defects in the plans and specifications cause defects in your work. You, as a contractor, are not required to guarantee the results of plans and specification. You merely follow those plans and specifications.

Macknight Flintic Stone Co. v. Mayor, 160 N.Y. 72 (1899), involved a contract for the construction of a courthouse and prison in New York City. The floor of the boiler room was approximately 26 feet below street level. The contract required Macknight to make the boiler room watertight. The contract included specifications detailing and itemizing the construction of the boiler room floor and walls, including a water-proof lining composed of six layers of roofing felt, solidly cemented together and coated with hot asphaltic cement. The court found that the materials and construction complied with the contract and specifications in every respect. There was evidence tending to show that a part of the materials required by the specifications were of an improper pattern. The court felt that Macknight had agreed to make the boiler room watertight by following the plans and specifications. Macknight did not guarantee those plans. The court found that the responsibility rests with the party who proposes the plan and represents to the contractor that it is adequate for the purpose to be accomplished. The court found that the work was performed faithfully according to the plans and specifications. The failure to secure a watertight boiler room was wholly owing to the defective design. See also *Dwyer v. Mayor*, 77 A.D. 224, 79 N.Y.S. 17 (1st Dept. 1902).

Ferrari v. Barleo Homes, Inc., 112 A.D.2d 137, 490 N.Y.S.2d 827 (2nd Dept. 1985), involved a dispute with respect to the construction of a home. The Ferraris hired a draftsman, who prepared the roof design. It was undisputed that the roof was constructed by Barleo according to the plans and specifications provided by the draftsman. Barleo would not be held responsible for damages resulting from the defective roof design.

As a contractor, you are smiling. However, a caution is in order. Where the contract contains an expressed guarantee, without qualification or exception, you must abide by that contract in spite of extra costs, which may be incurred due to defects in the specifications. *Cameron-Hawn Realty Co. v. Albany*, 207 N.Y. 377 (1913), involved a paving contract. The prices payable to Cameron-Hawn were not only for laying the pavement, but for maintaining it for a period of ten years. Cameron-Hawn expressly agreed to maintain the pavement through that period by making repairs and furnishing materials necessary to keep it in a specified condition. The pavement was not to show any wear greater than was usual with block pavement of the best quality under equally heavy traffic. The City of Albany was to retain five percent of the contract price in order to guarantee that Cameron-Hawn would adhere to the agreement.

Cameron-Hawn claimed that the contract fixed the size and dimension of the asphalt blocks with which it was to construct the pavement. The contract also specified the ingredients, quality, and proportion of the blocks' component parts.

The court noted that Cameron-Hawn had agreed to keep the pavement in a condition of repair and soundness for ten years. It distinguished *Macknight Flintic Stone* by stating that Macknight would only be applicable if the contract had provided that Cameron-Hawn would construct pavement, which would remain sound and perfect for a period of ten years by following the requirements and plans contained in the contract.

The court noted that the contract it was considering expressed an understanding between the parties that the pavement might show unsoundness or become defective. Under such a provision, Cameron-Hawn had the risk that the pavement would not remain sound and would require extensive repairs.

The moral of the story is that something that looks like a guarantee probably is. If a shadow of doubt crosses your mind, the provision should be discussed with your attorney.

§ 3.2.3 Requirements for Tests

If your contract requires that certain systems or materials are subject to a performance test, such a requirement will be upheld by the courts.

Logan v. Consolidated Gas Co., 107 A.D. 384, 95 N.Y.S. 163 (2nd Dept. 1905), involved the construction of a gas holder with steel tank. Upon completion of the tank, it was to be tested and proven watertight by being completely filled with water and remaining filled for 30 days. The tank was completed and the test began. Ten days into the test, the entire structure collapsed and was reduced to a scrap heap. Logan sought to recover his contract balance. The failure to meet the test defeated Logan's recovery.

Hagadorn v. McNair, 109 A.D. 759, 96 N.Y.S. 417 (4th Dept. 1905), involved the contract for drilling a well. The written agreement provided that Hagadorn would not charge McNair until the well had "been thoroughly tested and found to produce 20 barrels of water in each 24 hours." Hagadorn drilled the well to the depth of 52 feet. Upon testing, it failed to produce sufficient water.

The parties made subsequent oral agreements. Hagadorn returned to the site and drilled to a depth of 108 feet. Pursuant to the agreement, McNair was to test the well within 30 days. Although McNair had his pump put in place, he made no test within the 30 days. The test was not made until almost seven months later and evidence showed that the well did not produce 20 barrels of water. The court noted that, in general, Hagadorn would need to prove that he had fulfilled his agreement. However, McNair failed to accept or reject the well within 30 days. McNair had no right to postpone the test indefinitely. By not making the test, McNair never gave Hagadorn the opportunity to drill deeper. The court found for Hagadorn.

North-Eastern Const. Co. v. New York, 217 N.Y. 320 (1916), involved a contract for furnishing and installing mechanical and electrical equipment, including electric generating, distributing and lighting equipment, a boiler plant, steam piping, heating systems, and elevators at certain piers in the Hudson River. In connection with the electric generating and lighting provisions, North-Eastern agreed to furnish a competent crew to operate the plant for a period of six days in order that the proper test of the equipment be made. North-Eastern sought to recover for the cost of furnishing and delivering 306 tons of coal and the labor for firing the boilers in connection with the test. Having agreed to make the test, the court found that North-Eastern was required to furnish all materials and labor necessary to complete the test. Consequently, North-Eastern could not recover for the cost of the coal or the labor involved in the testing procedure.

A contract may specify tests of performance. The passing of those tests is deemed conclusive proof of performance. The court will find that the contractor had performed upon the completion or fulfillment of those tests. In *McIntosh Ready Mix Concrete Corp. v. R.D. Battaglini Corp.*, 36 A.D.2d 561, 317 N.Y.S.2d 692 (3rd Dept. 1971), McIntosh agreed to supply concrete to Battaglini for construction of a flood control project. The contract contained provisions that the concrete must pass certain slump (water cement ratio), air content, and compressive strength tests. The concrete passed the slump and air content tests. However, the reports issued by a testing laboratory indicated that the 28-day compressive strength test was below the contract minimum. There was ample evidence, however, that those who controlled the test sampling violated the standard testing requirements in various respects. There was, therefore, no justification in rejecting the concrete supplied by McIntosh. McIntosh was entitled to be paid.

The question of whether or not an apartment building has been completed was the issue in *20 East 74th Street, Inc. v. Minskoff*, 308 N.Y. 407 (1955). A co-op corporation was suing the promoters and developers of the co-op to recover damages for failure to complete the building according to contract. The co-op corporation sought $1,000,000 being held in escrow pending the completion of the building. The contract, however, contained three events by which completion of the building was to be determined. The first was the issuance of a Certificate of Occupancy by the City of New York. The Certificate of Occupancy would be granted by an affidavit "of a licensed architect, licensed professional engineer or superintendent of construction who supervised the construction". The Certificate of Occupancy must certify "that such a structure conforms substantially to the approved plans and specifications and the requirements of the laws governing building construction applicable to structures of the class and kind of such structure". The second event was that the final payment on the building loan mortgage had been made. The third event was that the mortgage had been assigned to the Mutual Benefit Life Insurance Company. The court pointed out that the three events depended upon performance of acts by third parties having interests adverse to Minskoff and almost identical with the co-op corporation. There was no suggestion that the performance of the three events was secured by fraud, collusion, or mistake. The court found that once the three events occurred, the building was deemed complete and, therefore, found for Minskoff.

§ 3.2.4 Standards Prescribed

Where the contractual provisions specifies a particular standard to be applied, the courts will uphold that application. *Early v. O'Brien*, 51 A.D. 569, 64 N.Y.S. 848 (1st Dept. 1900), involved a dispute concerning the construction of a building. The specifications provided that the basement floor was to be made perfectly watertight. As a contractor, your eyebrows should be raised at this. Even more seriously, although the specifications detail the method of construction, they also include "and must be warranted...". This would indicate that you, the contractor, are guaranteeing performance. The architect approved the method for construction of the basement floor and gave directions with regard to it. Not only that, but the architect was present during construction of the basement floor. The basement floor was never watertight. In fact, water split the floor right in the center and came flowing over it. The court found that the contractor was obligated to furnish a waterproof basement.

W.B. Armstrong Co. v. State, 186 A.D. 197, 173 N.Y.S. 786 (3rd Dept. 1919), involved a contract for the installation of heavy pipe for heating and domestic water supply mains at a state reformatory. It was conceded that the contract called for extra-heavy piping throughout. In the trade, extra-heavy piping possesses a thickness of 0.432 inch. Standard piping has a thickness of 0.280 inch. The contract provision in dispute read "*Extra Heavy Pipe.* - All extra heavy pipe shall be of full weight and thickness suited for working pressure of 250 pounds per square inch." Armstrong had begun installation of standard piping, arguing that it could withstand the working pressure of 250 pounds per square inch. The state objected and compelled use of extra-heavy piping.

The court rejected Armstrong's claim for the difference in cost for extra material and labor. The contract provision requiring that all extra heavy piping be "of full weight and thickness" mandated the use of piping possessing a thickness of 0.432 inch.

The dispute in *Raymond Concrete Pile Co. v. John Thatcher & Son*, 166 A.D. 522, 152 N.Y.S. 98 (2nd Dept. 1915), aff'd 221 N.Y. 715 (1917), centered about the provision that a concrete steel pile core "shall in each case be driven until not more than ten blows of a No. 2 Vulcan steam hammer are required to secure one inch penetration." In all, 213 piles were driven for Thatcher, who was building a mortuary vault in Greenwood Cemetery. As to 23 piles, Thatcher disputed performance in accordance with the contract. Raymond had stopped driving those piles while penetration of one inch was experienced with less than ten blows of the hammer. In some cases, driving had stopped at four, five, six, seven, eight, or nine blows.

The court found that the contractual standard of hammer blows was ten. If there were no penetrations of more than one inch under the last ten blows, then the pile driving could cease. The driving was to go on, however, until that standard was met. Raymond did not meet that standard for the 23 disputed piles and was not entitled to be paid for those piles.

The courts are strict in their interpretation of these contract standards. *Paramount Contracting Co. v. Deklade*, 169 N.Y.S. 154 (1st Dept. App. T. 1918), involved a contract for alterations to a saloon. Marble steps and risers were to be furnished and installed at the entrance. However, the contract only provided that the step was to be 1 1/4 inch thick. Paramount provided a step 1 1/4 inch thick, but the riser provided was only 7/8 of an inch thick. The court noted that the provision in the contract describing "steps and risers" had indicated that the thickness of the step alone was to be 1 1/4 inch. The provision was silent as to the thickness of the riser. Paramount was entitled to payment for the marble risers it provided.

§ 3.2.5 Workmanlike Manner

The courts generally require that a contractor perform his or her work in a skillful and workmanlike fashion. What are the boundaries of such a standard, and what obligations do they impose upon you as a contractor? The boundaries are extensive and the obligations great.

In *Mitchell v. Williams*, 80 A.D. 527, 80 N.Y.S. 864 (1st Dept. 1903), Mitchell contracted with Williams, an owner, to perform all carpentry work required in the erection and completion of a building. The contract stated that the work was to be done in a perfect and thoroughly workmanlike manner. The evidence showed that the material was not of that character or installed in the manner required by the contract. Recovery, therefore, was denied to Mitchell.

Van Deloo v. Moreland, 84 A.D.2d 871, 444 N.Y.S. 2d 744 (3rd Dept. 1981), also contained a contractual provision to do all work "in a workmanlike manner according to standard practices." Moreland had agreed to grade the site; install footings, piers, block wall and concrete slab; erect a pre-engineered building; and install insulation using materials supplied by Van Deloo. The court found ample support to show that the foundation was substandard, that one-half of the columns exceeded acceptable plumb standards and that the building was not square. Moreland had sued for the unpaid contract balance. In view of Moreland's breach of the contract, he was not entitled to recover that unpaid balance.

Is there an implied promise to perform a contact in a skillful and workmanlike manner? The court in *Fairbairn Lumber Corp. v. Telian*, 92 A.D.2d 683, 460 N.Y.S.2d 194 (3rd Dept. 1983), held that there is. Fairbairn had sued the owner on a contract to construct a log cabin with full basement and concrete septic system. The sidewall logs were improperly installed and not corked. Fairbairn had installed neither joint hangers on the beams nor bearing plates on foundation caps. The record demonstrated performance substandard to that generally accepted in the industry. Fairbairn did not collect its balance due and, in fact, the owner was awarded the expense of corrective work.

What exactly are the courts trying to accomplish here? A workmanlike standard is understandable where it is contractually agreed to. What occurs when that standard is elevated to an implied standard and, as we will now see, to an implied warranty? It appears that the courts are attempting to protect the "public" from a contractor in a better position to know. Where a person, a homeowner for example, is completely ignorant of construction techniques and considerations, it is the contractor who must bear the burden of bringing those matters to that person's attention. The principle "let the buyer beware" does not provide a contractor the protection it might be supposed to provide. These issues are particularly acute in subsurface condition situations, where a contractor's knowledge is presumed to be superior to that of a layman.

The Cacecis entered into a contract to purchase real property and to have a one-family home constructed. Four years after title closed, Mrs. Caceci noticed that the kitchen floor had dipped and moved away from the wall. In *Caceci v. DiCanio Const. Corp.*, 132 A.D.2d 591, 517 N.Y.S.2d 753 (2nd Dept. 1987), appeal granted 70 N.Y.2d 608, 521 N.Y.S.2d 224 (1987), aff'd 72 N.Y.2d 52, 530 N.Y.S.2d 771 (1988), the problem with the Cacecis' home was due to settlement of the foundation built on poor soil. Testimony and photographs substantiated that organic materials, such as trees, underlay and extended into the foundation. DiCanio denied knowledge that organic material underlay the foundation. A subcontractor who had excavated the land denied seeing or depositing any organic material. The photographs and testimony formed a sufficient basis to infer with probability that DiCanio knew that the house was being erected on poor soil, in violation of an implied warranty of workmanlike construction.

In affirming the lower court's decisions, the highest court of New York noted that over 25 states recognize some form of implied warranty of habitability or skillful construction in the connection of the sale of newly constructed homes, mentioning among others, Alabama, Arizona, California, Colorado, Connecticut, Florida, Illinois, Iowa, Maine, Michigan, Missouri, New Jersey, South Carolina, Vermont, Washington, and Wyoming.

The court noted that the parties were involved in the purchase of a home, which was contracted for sale prior to construction. They do not bargain as equals with respect to potential latent defects. The builder-vendor maintains a superior position and is the only one who can prevent the occurrence of major defects. The responsibility and liability is therefore placed on the party best able to prevent and bear the loss, the builder.

A Mississippi case reiterates the liability of the building contractor. In *George B. Gilmore Co. v. Garrett*, 582 So.2d 387 (Mississippi, 1991), the court found that the contractor should have known the possible existence of expanding soil at a building site. The owners had bought an empty lot and decided to build a ranch-style house. After construction, the house suffered from settlement problems. Settlement was caused by expansive "yazoo" soil lying 5 feet beneath the surface. There was testimony that geology maps had been available for almost 20 years showing the existence of yazoo soil in the county. The court found that the contractor is liable for negligent performance of a contractual duty. The court ruled that the contractor had a duty to warn the owners of the dangers of yazoo soil and the advisability of conducting a soils test. The court found that the duty existed even though it was not customary in the contracting trade. The contractor's compliance with the plans and specifications did not relieve him of liability. The Mississippi court found a duty to inspect the design for reasonably discoverable defects. A contractor's obligation in Mississippi is almost frightening.

J.R. Lewis v. Anchorage Asphalt Paving Co., 535 P.2d 1188 (Alaska 1975), involved a contract for paving certain streets in a trailer court. The contract contained the provision: "All work is to be completed in a workmanlike manner according to standard practices." The court found it well settled that a contractor, holding himself or herself out to be specially qualified to do a particular type of work, is subject to an implied warranty that the work would be done in a workmanlike manner.

Mr. Lewis had an engineering degree from a California college. He had worked for the design section of the California Highway Department. Upon purchasing the 40 acres of land, which he began to develop into a trailer court, he acted as his own general contractor. In leveling the trailer sites, he spread a glacial till of boulders, gravel, dirt, clay, some sand, and fine silt over the existing peat base. The upper layer of fill ranged from two to six feet and rested on a peat base of ten to twelve feet in thickness. Certain streets were paved by Anchorage. Defects and deficiencies began to appear some three to four weeks after completion. In three years, all the streets had begun to break up. At trial, Anchorage's position was that the term in its contract, "prepare subgrade," did not encompass the laying of crushed rock or gravel subgrade. It asserted that its asphalt was not defective. The court concluded that "the subgrade will be prepared by us" meant shaping and compacting of existing material. The court found that Anchorage had breached its contract to complete the work in a workmanlike fashion, by virtue of its failure to warn Lewis that a suitable result could not be obtained by placing paving over the glacial till material covering the peat.

The court found no authority that would place an affirmative duty on the owner to inform Anchorage that the fill was composed of glacial till covering a peat base.

To perform a contract in a workmanlike manner, a contractor is obligated to warn an owner of any potential defects in the construction project that come to the contractor's attention or should come to his or her attention due to his or her superior knowledge. See also *Wurst v. Pruyn*, 250 La. 1109, 202 So.2d 268 (Louisiana 1967).

Ridley Invest. Co. v. Croll, 56 Del. 208, 192 A.2d 925 (Delaware 1963), involved the construction of a post office building. The contractor complied with the plans and specifications furnished by the owner. Upon discovering soft soil during excavation of the site, the contractor notified the owner for the need of extra pilings for walls and floors. The owner authorized additional piling for walls only. The contractor was not responsible for repairing a floor which subsequently settled. The contractor had performed his work in a workmanlike fashion and was not liable for defects in the plans and specifications.

§ 3.2.6 Compliance with Law

A contractual provision requiring compliance with the laws or regulations will be upheld. *Bonagur v. Purificato*, 146 N.Y.S. 1070 (1st Dept. App. T. 1914), involved a contract for tearing down an existing dumbwaiter shaft and building a new one. Bonagur admitted that Purificato had told him there were violations in connection with the old shaft. The court found that Bonagur had held himself out as competent to erect a dumbwaiter shaft that would not be unlawful. Bonagur conceded that the law required the shaft to be of 4-inch blocks. Since Bonagur erected the shaft with 2-inch blocks, recovery was denied.

In *Reale v. Linder*, 135 Misc.2d 317, 514 N.Y.S.2d 1004 (Dist. Ct. Nassau Co. 1987), Reale sought to collect an unpaid balance on a home improvement contract.

Every home improvement contractor has an implied duty to perform the contract in accordance with fire prevention and building code requirements. A homeowner relies upon the contractor's skill and expertise in performing alternations and improvements. Reale constructed an addition with a crawl space of less than the full 18 inches required by law. Reale's failure to construct the addition in accordance with code requirements renders the addition illegal and frustrates the purpose of the contract. Not only could Reale not recover for the contract balance, but the court awarded Linder the costs for his corrective work. See also *Brainard v. Ten Eyck*, 102 Misc. 20, 168 N.Y.S. 116 (Albany Co. 1917).

3.3 SUBSTANTIAL PERFORMANCE

§ 3.3.1 Rationale for Recovery

The question of substantial performance often arises in construction contracts. The courts will not excuse a deviation from plans and drawings. However, the courts will say that a trivial deviation does not result in a breach of contract with forfeiture of the contract price, but rather in a credit for any resulting damages.

The touchstone case supporting the theory of substantial performance is *Jacob & Youngs, Inc. v. Kent*, 230 N.Y. 239 (1921), reh'g denied 230 N.Y. 656 (1921). Jacob & Youngs sought to collect their contract balance from Kent, owner of a country residence, which they had constructed. One of the plumbing specifications provided: "all wrought iron pipe must be well galvanized, lap welded pipe of a grade known as 'standard pipe' of Reading manufacture." Almost a year after Kent took possession of this home it was discovered that some of the plumbing pipe was the product of other factories. Kent's architect directed that the work be removed and replaced.

The court noted that the plumbing was encased within the walls of the house except in a few places where it was exposed. Compliance with architect's order would mean more than substitution of other pipe. Compliance would have involved demolition, at great expense, of substantial portions of the structure. The omission of the prescribed brand of pipe was neither fraudulent nor willful. Reading pipe is distinguished from Cohoes Pipe and other brands only by the name of the manufacturer stamped upon it at intervals of between 6 and 7 feet. Kent's architect, though he inspected the pipe, failed to notice the discrepancy.

The courts never say that one who makes a contract fulfills his or her duty by less than full performance. However, an omission both trivial and innocent will not result in a forfeiture of the contract by breach. See *Bowen v. Kimball*, 203 Mass. 364. The court determined that the contract is not to be forfeited for breach. Allowances were to be made, not for the cost of replacement, which would be great; but for the difference in value, which would be either nominal or nothing. In most cases, the cost of replacement is the measure. The owner is entitled to money, which will permit him to complete, except where the cost of completion is grossly or unfairly out of proportion to that to be obtained. When that is true, the measure is the difference in value. See *Foeller v. Heintz*, 137 Wisc. 169. Jacob & Youngs prevailed.

The question of substantial performance often turns on the facts. In *Glacius v. Black*, 50 N.Y. 145 (1872), admittedly a pre-*Jacob & Youngs* case, George Glacius was to perform certain alterations to the home of Bessie Black. The work was principally taking off an old peak roof and substituting a mansard roof, finishing off the upper story, and making other repairs to the interior of the house. Although the court's decision was based on the incapacity of the owner's architect to waive certain requirements, the court did note that whether defects are substantial or technical is a question of fact.

While it is next to impossible for a builder to comply literally with minute specifications of a building contract, a court will consider the extent of defects and deviations from the specifications. The court found that the defects here were quite numerous. There were defects in plastering, the walls were out of plumb, the windows were too narrow. Glacius failed to place sills under cross partitions. There was leakage in the top floor. Folding doors were too narrow and too short. Lightning rods torn down were not replaced. Defective materials and workmanship were used in the moldings and casings. The court found that the architect could not waive the requirement that these items be performed correctly. The court notes that where defects and omissions are so numerous and pervasive as to show that the job was done in a slovenly and improper manner, without conforming substantially to the plans and specifications, then the contractor is not entitled to compensation.

The extent of nonperformance will determine whether the theory of substantial performance is applicable. George Spence agreed to erect a building for Albert Ham. *Spence v. Ham*, 163 N.Y. 220 (1900). Although there were certain minor omissions, the court also found that there was a failure to place bridging in certain places. Certain collar braces were omitted. Girders were not of the specified length and were not properly placed. Trimmers and headers were single rather than double. The minor deviations were a failure to put drawers and shelves in closets and to place a wooden partition on a brick wall in the basement.

The court found that the law will not be satisfied by allowing the expense of a new girder. A defective girder, partially supporting a building, must be removed and another put in its place to remedy the defect. While it may be possible to make the substitution, the process is difficult and apt to injure the structure. The court finds for the owner, since this cannot be considered an unsubstantial defect.

§ 3.3.2 Building Contracts

Because of the difficulty in complying with each minute specification in a building contract, the courts have devised the standard of substantial performance. Substantial performance of a contract will support a recovery for the contract price less allowances for any minor deficiencies. *Manshul Const. Corp. v. D.A.*, 111 Misc. 2nd 209, 444 N.Y.S. 2nd 792 (Kings Co. 1981), involved the construction of a physical education building at Kingsborough Community College. There is substantial performance of a construction contract where all essentials necessary for the full accomplishment of the purpose for which the building has been constructed are performed. The owner should obtain substantially what is called for in the contract. The Dormitory Authority had claimed that certain items of work had not been performed by Manshul. However, it presented no evidence to indicate any such deficiencies. The court found that Manshul substantially performed.

How does substantial performance affect you, the contractor? *Pilgrim Homes & Garages, Inc. v. Fiore*, 75 A.D. 2nd 846, 427 N.Y.S. 2nd 851 (2nd Dept. 1980), appeal dismissed 51 N.Y. 2nd 702, 431 N.Y.S. 2nd 1030 (1980), involved a contractor's attempt to recover an outstanding balance for constructing a luxurious one-family prefabricated home. There is no question that Pilgrim did not fully perform. The court notes, however, that the failure to establish full performance does not preclude Pilgrim's right to recover. The question is whether the Fiores have essentially received what they bargained for: a prefabricated luxury home. The court notes that many of the items of dispute contain modifications and additions made at the insistence of the Fiores as part of their incessant efforts to customize the prefabricated home. The court concluded that Pilgrim substantially conformed to the construction contract. It noted that neither the absolute number of incompletely or defectively performed items nor the relative percentage of the value of such items to the total work is dispositive. A consequence of the holding that Pilgrim substantially completed is that Pilgrim is necessarily precluded from recovering the full contract price. Pilgrim's recovery is limited to an adjustment for the items not duly performed. However, the burden of proving the cost of those items omitted is on the contractor as a condition of recovering on the theory of substantial performance. The court provided examples of items to be considered in determining the decreased value due to a defect or omission. Among those items were a damp basement and cracks in the foundation, the fact that bay windows were supported by concrete sheets rather than a fully poured foundation, a failure to provide the requisite concrete foundation under the front porch, a failure to double-joist the rafters, the alleged improper installation of a steel support beam, issues concerning the smoking of a fireplace and a chimney constructed as to allow only a 16-inch closet in one of the bedrooms. In addition, technical violations of the building code should be considered as well as a failure to properly grade the floor surrounding a swimming pool, the installation of a defective stairway bannister and the failure to install paneling in the library.

From the length of the list it is obvious that substantial performance may leave a considerable amount of work unperformed. Remember, the standard set by this court was whether the homeowners essentially received their bargained-for result.

§ 3.3.3 What Constitutes Substantial Performance

As the court noted in *Pilgrim Homes*, in connection with substantial performance, two questions are asked. What is the bargain? Has that bargain been received except for minor or unimportant deviations? In *Schultz v. Goodstein* 180 N.Y. 248 (1905) the contractor for plumbing work and materials in a block of five dwellings furnished earthen sewer pipes, flush pipes of "E" thickness, and waste pipes 1 1/2 inches in diameter. The contract required iron sewer pipes, flush of "D" thickness and waste pipes 2 inches in diameter. The evidence showed that these deviations were not inadvertent or unintentional. They were a willful and intentional departure from the contract.

The contractor is to fairly perform according to the contract terms. Any substantial change is made at his or her risk. In order to avoid injustice, the law will accept unsubstantial deviations made in good faith. Where substantial completion is achieved, the law will permit recovery only after the appropriate deductions have been made. The contractor has no right to substitute his or her own judgment for the provisions of the contract. Where an owner stipulates iron pipe, he or she has a right to iron pipe. The contractor could not recover his contract balance.

The application of the doctrine of substantial performance is considered in *Cramer v. Esswein*, 220 A.D. 10, 220 N.Y.S. 634 (2nd Dept. 1927). Cramer was to install eight radiators. He installed only seven, leaving a bathroom without heat. In contravention of his contract, he installed a second-hand bathtub and wash basin. In addition, he willfully removed doors and other fixtures from the furnace. The doctrine of substantial performance has no application where there is an intentional, deliberate, and willful departure from the contract.

Each case is different; and courts must consider the individual facts of each case. Substantial performance is not a black and white term. There is not a specific percentage that indicates substantial performance. However, cases have looked at percentages completed in determining whether the performance has been substantial. In *Gompert v. Healy*, 149 A.D. 198, 133 N.Y.S. 689 (2nd Dept. 1912), the court held that damages of nearly 25 percent of the contract amount, for failure to perform, would prevent recovery based on substantial performance.

Cassino v. Yacevich, 261 A.D. 685, 27 N.Y.S.2nd 95 (3rd Dept. 1941), concerned a contract for Margaret Cassino to build a one-story brick garage with wood frame living quarters. The total contract amount was $5,600.00. It was found that 22 items of Ms. Cassino's work were improperly performed amounting to $1,372.00 or slightly over 24.5 percent of the entire contract price. That amount was credited to the owner. The appellate court found that this could not be substantial performance and that Ms. Cassino could not collect her contract balance. The court noted that substantial performance has been a fruitful source of litigation. The court reiterated that slight and insignificant imperfections or deviations may be overlooked, but the contract must in all other respects be performed according to its terms. The court cited cases in which the doctrine of substantial performance did not apply where the amount uncompleted ranged from 10 percent to 25 percent. Perhaps the most severe case is in *Bowen v. Kimball*, 203 Mass. 364, where the contractor was denied recovery where a deviation diminished the value of the building by less than 1 percent of the total contract price. Massachusetts case law notwithstanding, the moral of the story appears to be that for the doctrine of substantial performance to apply, the contract must be reasonably performed. When a full 10 percent of the contract is not, absent mitigating circumstances, a court will generally not find substantial performance. However, as explained in connection with *Pilgrim Homes*, the percentage complete is not the sole determining criterion. The entire relationship of the parties must be considered.

§ 3.3.4 Deviations

Deviations must be inadvertent and unintentional. Willful, intentional, and deliberate departures from the contract as in *Schultz v. Goodstein* will prevent application of the doctrine of substantial performance.

3.4 EXCUSES FOR NONPERFORMANCE

§ 3.4.1 Failure of Consideration

What exactly is failure of consideration? Even more basically, what is a want of consideration? A failure of consideration occurs after a contract has been made. Some supervening cause or event alters the rights of the parties. The event is not within the contemplation of the parties at the time of contract. Where the consideration for the promise fails, and a party gets nothing for his or her bargain, then the promise cannot be enforced.

Want of consideration is simply this. For a promise to be binding, there must be an exchange. Consideration is the object exchanged for the promise. In a construction contract, it is a simple enough matter. For you, the contractor, consideration for your promise to construct a building is the receipt of the contract price from the owner. In exchange for your promise, the owner suffers a legal detriment. He or she obligates himself or herself to pay the contract price to you.

Can the argument of a want of consideration be raised in a construction contract? The argument was raised in *Young Foundation Corp. v. A.E. Ottaviano, Inc.*, 29 Misc.2nd 302, 216 N.Y.S.2nd 448 (West. Co. 1961), aff'd 15 A.D.2nd 517, 222 N.Y.S.2nd 685 (2nd Dept. 1961). Ottaviano was a prime contractor with the State of New York for construction of a portion of the Palisades Interstate Parkway. Ottaviano entered into a written subcontract with Young by which Young was to furnish and install H-beam steel piles and other equipment. Young sought its contract balance due.

Young was unable to buy the steel piles from Bethlehem Steel Corporation because Bethlehem refused to extend credit to Young. Ottaviano orally agreed to guarantee to Bethlehem the payment of steel support for the job. Young orally agreed that Ottaviano might deduct the purchase price of the steel from the amount due under the subcontract. Ottaviano notified Bethlehem of the guarantee. He informed Bethlehem that it might either bill Ottaviano directly or bill Young. Bethlehem billed Ottaviano directly. Ottaviano claimed that the oral modification was without consideration and, therefore, not binding.

Young was plainly obligated before the oral obligation to both furnish and install the steel piles. In this context, Ottaviano's promise to butress Young's credit seems without consideration. A promise made to induce a party to perform what he or she has contracted to do is not supported by consideration. But this rule applies only to those contracts that have not been performed. A party cannot defend against a performed contract on the grounds of want of consideration. The court rejected Ottaviano's argument. Young received its contract balance.

§ 3.4.2 Difficulty

Can you be released from your contract because of dificulty in performing that contract? Does the probability of great financial loss allow you to be released?

The fact that performance becomes more difficult or more expensive than contemplated does not justify nonperformance. When two parties bind themselves to a contract, the courts will not alter that contract because of hardship. *Standard Oil of New York v. Central Dredging Company*, 225 A.D. 407, 233 N.Y.S. 279 (3rd Dept. 1929), aff'd 252 N.Y. 545 (1929), involved an agreement by Central Dredging to dredge 40,000 cubic yards of soil from the Hudson River and deposit it on Standard Oil's premises on Van Rensselaer Island. It was also to deposit thereon 100,000 cubic yards of spoil from a government contract on which it was working.

Pursuant to its contract with Standard Oil, Central was to apply for and obtain all necessary permits for town, municipal, state, or federal authorities. The federal government has authority over the waters of the Hudson River. It was, therefore, necessary to obtain a federal permit. The permit was refused by the United States division engineer of the Corps of Engineers. The determination of the division engineer could have been reviewed by the chief engineer and, in his turn, by the Secretary of War.

Central Dredging argued that the refusal by the division engineer abrogated its contract with Standard Oil. Central cited the rule that when continued existence is assumed and is the basis of an agreement, destruction or unavailability puts an end to the obligation. The court held that Central failed to show sufficient diligence in its efforts to obtain a permit. Difficulty or even improbability of accomplishment without great financial loss will not release a contractor. To establish the appropriate defense, the contractor would need to show that the thing contracted for cannot be accomplished by any means. Central Dredging was held to the contract. Standard Oil prevailed.

Kinser Const. Co. v. State, 125 N.Y.S. 46 (Ct. Cl. 1910), aff'd 145 A.D. 41, 129 N.Y.S. 567 (3rd Dept. 1911), aff'd 204 N.Y. 381 (1912), involved the construction of 3.76 miles of the New York Barge Canal System. While Kinser was excavating for one of the locks an extensive cave-in occurred, revealing that the balance of the earth was of a "slippery, greasy clay." Kinser's expert stated he had never seen soil like it and that it would not be good engineering practice to build a lock in such material. The state issued a stop order for six months. It then submitted to Kinser an alteration order involving extensive changes in the contract for the remainder of the work. Kinser refused to accept this alteration, insisting that it amounted to a fundamental change in the contract and a breach by the state. The state advertised for bids and completed the work without the contractor. A rule found in English jurisprudence as early as 1178 held that the inability to execute a contract due to subsequent unforeseen accident or misfortune without the fault of either party would not excuse performance. Where a party by his or her own contract creates a duty, he or she is bound to make it good. Accident or inevitable necessity does not excuse performance because the parties could have provided against such in the contract.

Are there exceptions to this rule? The law has changed since 1178. A party is excused where impossibility arises from a change in the law. A party is also excused where the specific thing essential for performance is destroyed. In a contract for personal services sickness or death will excuse performance. Finally, and applicable to this case, where conditions essential to performance do not exist, performance is excused.

In the course of construction of the canal, natural conditions of soil unexpectedly appeared. The contingency provisions in the contract did not, in expressed terms, cover the developments. The performance of the contract as planned was impossible. The law will read into the contract an implied condition, when the contract was made, that such a contingency will terminate the entire contract; a fine decision, but it should not instill a great deal of confidence. The art of contract of drafting has also advanced somewhat since 1912 and a public contract will cover just about any contingency imaginable. Public contracts should be read carefully and should be read by your attorney.

Finally, may a contractor recover additional costs incurred because the contract has become more expensive to perform? You, of course, know the answer to this question but let me give you a case example. In *American Store Equipment & Const. Corp. v. Buffalo Municipal Housing Authority*, 202 Misc. 222, 111 N.Y.S.2d 688 (Erie Co. 1952), aff'd 282 A.D. 824, 122 N.Y.S.2d 533 (4th Dept. 1953), aff'd 306 N.Y. 773 (1954), American entered into a contract for the construction of a wartime housing project in Buffalo. During the course of construction the President of the United States issued an Executive Order, which increased the minimum work week from 40 to 48 hours. American sought to collect increased labor costs due to overtime payments for the extra eight hours per week. The contract specifically stated that, "no labor, workman or mechanic...shall be permitted more than eight (8) hours in any one calendar day, nor more than five (5) days in one week..." except when otherwise proclaimed by the President of the United States. The court found that the parties did contemplate such a proclamation by the President. The court notes that even if the parties had not so contemplated and contracted for this contingency, there could be no recovery. Merely because the contract has become increasingly difficult and expensive to perform due to some law enacted after its execution, performance is not excused. The contractor's claim fails.

3.5 APPROVAL BY THIRD PERSON

§ 3.5.1 Sufficiency of Certificate

Just as the owner or the owner's agent may need, pursuant to the contract, to be satisfied by the work, the owner may also select a third party to determine the completion and sufficiency of the work involved. That third party is usually an architect or engineer. A provision that appoints that architect or engineer is generally valid.

In appointing the architect or engineer to determine the completion or sufficiency of the work, the provision may also require that that architect or engineer issue a Certificate of Completion. Where the provision requires that final payment may not be demanded without a Certificate of Completion, such certificate is a condition precedent to recovery. If your contract contains a provision regarding a Certificate of Completion, it should be reviewed by your attorney so that you may understand the obligations you are about to assume.

The courts are not insensitive to the realities and personality conflicts that may be involved in a construction contract. The fact that the requirement for a certificate of completion is a condition precedent does not absolutely bar recovery. In *Crouch v. Gutman*, 134 N.Y. 45 (1892), a request was made for the architect's certificate. The architect refused to grant the certificate, stating that many of the items were defective. The court noted that at the time the request for the architect's certificate the work had been done and defects in the work had been removed. Minor variations could have been made to conform with a small expenditure. Where the work is substantially performed, the refusal by the architect to grant the certificate was unreasonable. The nonproduction of the certificate is excused.

National Contracting Co. v. Hudson River Water Power Co., 192 N.Y. 209 (1908), involved a contract for construction of a masonry dam. Midway through the project the water power company proposed that the construction of the dam be changed to substitute a dam of earth with a masonry core. National protested and, after giving notice, withdrew from the site and ceased prosecution of the work.

A provision in the contract stated that all questions that might arise as to the amount or quantity of work to be paid for were to be decided by the engineers of the water power company. The company argued that since the decision of the engineers had not been sought, the action could not proceed. The court found that prosecution of the action was not barred. The action was brought to recover a balance due for work done under the contract and for damages for breach. The contractual provision did not provide that all controversies were to be subject to the decision of the engineers. The court notes that the certificate of an engineer, unless unreasonably refused, is a necessary prerequisite for recovery for work done. However, the absence of such a certificate does not bar this action for damages caused by the company's breach of contract.

What form must a Certificate of Completion take? Wyckoff & Winham, carpenters and builders, entered into a contract with Meyers to construct three brick stores in New York City. *Wyckoff v. Meyers*, 44 N.Y. 143 (1870). The contract provided that Wyckoff & Winham obtain a certificate from the architects to the effect that the work was completely finished. Meyers resisted payment arguing that the certificate did not specifically certify that "the work was completely finished."

The court found that the certificate substantially contained all that the contract required. The architects had certified that the last payment was due. The court found that this was clearly the same as if the architects had certified that the work was completely finished. Wyckoff & Winham were entitled to their last payment.

In *Snaith v. Smith*, 7 Misc. 37, 27 N.Y.S. 379 (N.Y. Co. 1894) the contract required that final payment be made upon the architect's certification that the work was done to his satisfaction. The architect did issue a certification that the final payment was due. Smith resisted payment. The court noted that no fraud, mistake, or bad faith by architect in providing the certificate was either proved or attempted to be proved at trial. In absence of such proof, the granting of the certificate was conclusive on the owner. The certificate stating that, "there is now due...final payment on this contract...." was sufficient.

The building contract in *Strauss v. Hanover Realty & Constr. Co.*, 133 A.D. 743, 118 N.Y.S. 193 (1st Dept. 1909), provided that only the final certificate of the architect would be conclusive evidence of performance of the contract. Two letters had been written by the architect, advising the contractor to make certain corrections. The court rejected the argument that these letters constituted a certificate of final payment.

The court found that the letters did not constitute a final certificate within the meaning of the contract. The letters did not purport to be certificates. They did not indicate a final and thorough examination of the work by the architect. Finally, the letters did not indicate that the architect found the work satisfactory according to the terms of the contract.

§ 3.5.2 Conclusiveness

The courts will enforce a provision by which the parties select a person to determine questions with regard to the contract. When the parties stipulate that that person's determination is final and conclusive, the parties are bound by the determination.

The person selected by the parties in *Sweet v. Morrison*, 116 N.Y. 19 (1889), reh'g denied 23 N.E. 897, reh'g denied 24 N.E. 1097, was the engineer in chief for the building of a railroad extending from the Missouri River to the Pacific Ocean. Morrison had a contract with the Northern Pacific Railroad Company to build a part of its road extending from the Missouri River to the Red River, consisting of more than 200 miles. Morrison subcontracted with Sweet for the construction of bridges, trestles, and other timber work. In the subcontract, Sweet agreed to perform to the satisfaction and acceptance of the chief engineer. Both parties agreed that to prevent disputes and misunderstandings the chief engineer of the Northern Pacific Railroad was to decide all matters arising or growing out of the subcontract. The decision of the chief engineer on any matter touching on the subcontract was to be final and conclusive between the parties. The parties waived any right to actions, suits, or other remedies at law. A dispute arose as to the amount of work performed by Sweet. It was decided by the chief engineer. The court noted that the person selected by the parties to make the determination was in employ of neither. As chief engineer of the Northern Pacific Railroad, he sustained a relationship to both of them.

This case was unlike those frequently arising in which the certificate is required from an architect or engineer in the employ of one of the parties. In those cases, there is a danger that the person acting as arbitrator might favor his or her employer. This legal disability will not hinder such a person from being an arbitrator, provided that the fact of his or her employment is known to the parties at the time of the agreement. However, impartiality is a fundamental requisite. The courts will closely scrutinize the action of an arbitrator whose relationship to one of the parties might naturally influence his or her judgment.

In this case, however, there was no reason why the chief engineer should not be wholly disinterested. The determination made by the chief engineer should not be set aside or disregarded unless some good reason is shown for such an action. It was the intention of the parties to make the chief engineer's determination final. Absent proof of corruption, bad faith, misconduct or palpable mistake appearing on the face of the determination, neither party can prove that the chief engineer's decision was incorrect.

Elizabeth Sash, Door & Supply Co. v. St. Vincent's Hospital, 241 A.D. 751, 270 N.Y.S. 195 (2nd Dept. 1934), involved a dispute regarding a final determination by an architect. The court noted that an architect must act in a judicial manner, not hastily or capriciously. He must apply reasonable care and good judgment in making his judgment. The court found that the architect's decision was not based on whim, caprice, or neglect. The architect's decision was upheld by the court.

The contract in *Chas. S. Wood & Co. v. Alvord & Swift*, 232 A.D. 603, 251 N.Y.S. 35 (1st. Dept 1931), aff'd 258 N.Y. 611 (1932), was for the construction of the New

York Central building. James Stewart & Company were the general contractors. They entered into a subcontract with Alvord & Swift for the installation of heating and ventilating equipment. The subcontract included furnishing a nonconducting covering for the piping. Alvord & Swift then subcontracted with Wood for the pipe covering. Wood sought a balance claim to be due under its subcontract. Alvord & Swift resisted payment claiming that Wood failed to complete the work and that 25,000 feet of radiator connections had not been covered. Alvord & Swift claimed a credit for the value of the work omitted.

The subcontract specifications stated, "All high pressure, reduced pressure, low pressure, high and low pressure, drip piping, all concealed heating risers and radiator connections, steam return, drip and vapor piping shall be covered as hereinafter specified." Wood claimed that the word "concealed" related not only to "heating risers" but also to "radiator connections." Alvord & Swift claimed that term was limited in its application to "heating risers" and that Wood was obligated to cover all radiator connections whether concealed or exposed. The subcontract between James Stewart and Company and Alvord & Swift stated that the architects shall interpret the drawings and their interpretation as to the meaning thereof shall be accepted as final and conclusive. The clause was incorporated by reference into the subcontract between Wood and Alvord & Swift. Wood requested a ruling from the architects who stated that the heating and ventilating system was entirely satisfactory and conformed to the plans and specifications. The court noted that the interpretation by the architect is controlling in the absence of fraud or bad faith. While Alvord & Swift disagrees with the conclusion of the architect, such disagreement presents no triable issue. Wood was awarded the full amount of his subcontract. See also *Callaghan v. City*, 233 A.D. 263, 251 N.Y.S. 501 (1st Dept. 1931).

In *Helmer-Cronin Constr., Inc. v. Central School District*, 51 A.D.2d 1085, 381 N.Y.S.2d 347 (3rd Dept. 1976), a dispute arose during the course of a construction of a new elementary school. Was material excavated at the construction site satisfactory for use as backfill? The architect decided that the excavated material was suitable and disapproved the use of any offsite material. Helmer disagreed and proceeded to use some offsite fill. School District refused to pay for the offsite fill.

Under the terms of the contract, the architect was made the interpreter of the contract and authorized to make all determinations regarding the acceptability of the work and materials. Helmer was required to provide all necessary materials and guarantee them against defects for one year. The contract provides that fill and backfill material may be material from the excavation, if approved by the architect, and if it meets the compaction requirements.

The courts notes that it is well settled that any decision of the architect within his provenance is conclusive and binding absent a showing of fraud, bad faith, or palpable mistake. The contract, however, provided that not only must the onsite material used for backfill be approved by the architect, but was also conditioned upon that material meeting specified compaction requirements. The court found, therefore, that the parties intended to withhold from the architect the absolute power to determine the suitability of onsite fill.

There was evidence that the onsite fill did not meet specified compaction requirements. The material was analyzed by a testing company selected by the architect and that company recommended that the onsite, excavated material not be used as backfill. Helmer's independent laboratory supported this conclusion. The court found that the architect's disapproving the use of offsite material could not be sustained as a matter of law. The issue of bad faith or palpable mistake on the part of the architect was put at issue by the evidence.

Naturally, the courts will carefully examine the scope of a contractual provision granting a third party the power to make final and conclusive determinations. *Application of Liebhafsky*, 100 A.D.2d 764, 474 N.Y.S.2d 723 (1st Dept. 1984), aff'd *Liebhafsky v. Comstruct Associates*, 62 N.Y.2d 439, 478 N.Y.S.2d 252 (1984), involved a contract for renovations to a Manhattan townhouse. The contract provided that claims and disputes relating to the execution or progress of the work or interpretation of the contract document shall be referred initially to the architect for decision. The contract also provided that matters submitted to the architect are subject to arbitration upon written demand of either party made on or after the date on which the architect rendered his decision or the 10th day after which the parties presented their evidence to the architect. The contractor, Comstruct Associates, Inc., unilaterally issued approximately 20 change orders related to work being done at the townhouse. Subsequent to the issuance of the change orders, Liebhafsky complained stairway bracing did not pass inspection, that there was inadequate onsite supervision, and insufficient recordkeeping. Liebhafsky terminated the contract pursuant to one of the general conditions. Comstruct served a demand for arbitration.

In interpreting similar conditions, the courts had held that a claim for damages, caused by a substantial delay, asserted approximately 20 months after substantial completion did not fall within the scope of disputes requiring reference to the architect. In this situation, although there was no substantial completion, Liebhafsky had terminated the contract and Comstruct had ceased construction before any demand for arbitration was made. The court found that the scope of the architect's authority is limited to, "the operational phases of the construction." Comstruct's claim was not subject to review by the architect.

Comstruct's claims were for money due on the contract and did not relate to execution or progress of the work. Nor did they relate to interpretation of contract documents. Comstruct's claims were subject to the arbitration provisions and thus should proceed directly to arbitration. Referral of the claim to the architect was not a condition precedent to arbitration.

§ 3.5.3 Particular Determinations

As we have seen, the determination of an architect or engineer authorized by both parties is conclusive, absent proof of corruption, bad faith, or palpable mistake. In what factual circumstances has the architect's power been exercised?

In *Joseph Davis, Inc. v. Merritt-Chapman & Scott Corp.*, 27 A.D.2d 114, 276 N.Y.S.2d 479 (4th Dept. 1967), later appeal 30 A.D.2d 765, 291 N.Y.S.2d 839 (4th Dept. 1968), we are shown clearly the distinction between what the architect or engineer can determine and what must be left to the court. Merritt and the Power Authority of New York State contracted for construction of a power plant. Merritt entered into a subcontract with Davis, whereby Davis was to perform the mechanical piping system work.

The contract stated that any ambiguity in the plans or specifications must be submitted to the Authority for adjustment. The Authority's determination would be final and conclusive. The subcontract stated that any determination between Merritt and the Power Authority would be binding on the parties to the subcontract. The initial dispute involved Davis' claim for the installation of pipe supports for embedded piping and for the material and fabrication of the supports. The claim was submitted to the Power Authority, which asserted that the Power Authority was only bound to furnish material when piping was permanently mounted on hangers and supports. Where piping was to be embedded in concrete, all operations were to be performed by the contractor with the cost included in the bid price. Merritt's defense to Davis' claim was that the prime contract provided that any ambiguity in the specifications would be final and conclusive upon the parties. The subcontract, then, required that the decision between Merritt and the Power Authority shall be binding upon the parties to the subcontract. The court found that the language of the contract specifications was clear. The court noted that a contract provision, which makes the decision of a third party final, is conclusive absent fraud, bad faith, or palpable mistake. However, the legal meaning of a subcontract is always the responsibility of the court and not of the third party. The court agreed with Davis' interpretation and supported Davis' claim.

On the other hand, Davis asserted a claim for the installation of duriron pipe. The court recognized that the distinction between factual and legal questions and properly determined that the engineers were making a factual determination. The question of fact was whether or not it was more difficult to work with duriron pipe than with ordinary cast iron pipe. There was no claim of fraud or bad faith on the part of the engineer. The court found that this was an issue involving quality or classification and therefore one left to the engineer and not the court for disposition. The court will defer to the engineer's expertise.

In *Atlantic, Gulf & Pacific Co. v. Woodmere Realty Co.*, 156 A.D. 351, 142 N.Y.S. 953 (2nd Dept. 1913), appeal dismissed 209 N.Y. 557 (1913), the dispute involved estimates of quantities in the filling and grading of certain lowlands. The court noted that the estimates were not shown to be wrong or issued under any mistake. Those estimate determinations would be upheld unless shown to have been obtained by fraud, deceit, or palpable error. Nine certificates by the engineer were before the court showing a quantity dredged and deposited. The court found that evidence of a quantity lost through subsidence was too uncertain to afford a basis for disturbing the certification of the engineer.

Burke v. Mayor, etc., of New York, 7 A.D. 128, 40 N.Y.S. 81 (1st Dept. 1896), involved an engineer's determination regarding the amount of paving work installed.

The contract provided that a final certificate by the engineer was conclusive as to the amount of materials furnished and the work done. The engineer had issued the final certificate. Where there was no proof of fraud, there could be no recovery. However, where an engineer has, based on an erroneous construction of the contract, deliberately excluded work that was done by the contractor, the contractor may prove that he or she has done such work, that it is included in the terms of the contract, and that it ought to have been included in the amount of the engineer's certificate.

The court notes that there was no claim that the contractor did not do the work for which he sought compensation. The claim by the City was that it was outside the terms of the contract. Upon this theory, the engineer excluded the work from his final certificate. It is undoubtedly true that where there is a dispute in regard to work or its character, the certificate of the engineer is conclusive. Here, where the contract calls for performance of the work that the contractor has done, and the engineer, upon an erroneous construction of the contract has excluded it, the contractor has a right to recover. The court found that under these circumstances, the engineer has no right to reject the work as not being included in the contract.

A curious twist is placed on the final certification by the engineer in *Condict v. Onward Const. Co.*, 210 N.Y. 88 (1913). There, Condict constructed a refrigerating plant in an apartment building. The consulting engineer, as authorized by the contract, had given Condict a final certificate that the contract had been entirely completed and all defects corrected. The owner offered evidence that the plant, subsequent to the engineer's certification did not operate with the stipulated efficiency. Condict used the engineer's certificate as a shield and argued that it was a conclusive determination that the contract had been completely performed and all outstanding issues settled. While the certificate of a consulting engineer is conclusive proof in the absence of corruption, fraud, or palpable mistake, there was no hostility between a certificate of final completion and a warranty. The contract contained a specific guarantee as to the capacity and efficiency of the refrigerating plant after completion. The certificate of final completion secured the owner in the result that the plan should be constructed and delivered in substantial accord with the contract requirements. The express warranty secured the owner in the result that the plant should subsequently operate with a stipulated efficiency. The right of the owner to recover damages for breach of the express warranty survived the delivery and acceptance of the completed plant.

§ 3.5.4 Unreasonable Refusal

Should a contract specify that an engineer's certificate is required as a condition to payment, the courts will enforce that requirement. Suppose the architect unreasonably refuses to issue the certificate? The courts have regarded an unreasonable refusal as bad faith and will allow payment without the certificate. Procurement of the certificate is excused.

In *Ross v. New York*, 85 A.D. 611, 82 N.Y.S. 920 (2nd Dept. 1903), the city resisted payment to a contractor, James Gillen, who had installed an extensive sewer system. The contract required a certification that the work had been fully completed and materials fully furnished. The City argued that in the annular connection or joint between the pipes, cement was installed only in the upper portion of the joint and not all the way around, more especially down to and at the bottom of the pipe. In addition, jute was not sufficiently used. The sewer pipe was not round, but pear shaped, the broadest part at top and the pointed part at bottom. The spigot was only 1/4 or 3/8 inch smaller than the hub, and it was expected to be driven or butted home so as to fit closely. The court noted that after being thus laid, it would be difficult to drive jute or cement into the aperture without breaking the pipe. The City had concluded that the sewer pipe was improperly installed and brought in a contractor to perform corrective work. That contractor destroyed the entire length of sewer pipe in dispute, removed it and replaced it. During such removal it was noted that in a number of places there was no indication that jute and cement had been used at the bottom of the joints. There was evidence that jute was used whenever it was practicable to use it and the joints were properly cemented.

The court noted that there was a large flow of natural water from springs along the line of the sewer. Such a flow could have caused, after some time, a removal of the cement and destruction of the closed joint. Gillen's sewer pipes were laid under the supervision of an engineer from the sewer board, and any complaints made by the engineer were remedied by Gillen. When Gillen's pipes were destroyed and removed, there was evidence that they were not in perfect condition. However, there was evidence that Gillen had fully completed the work substantially in accord with the contract terms and conditions. The court found that the refusal of the City's sewer commissioners to certify completion of the work was unreasonable. Gillen was entitled to collect its contract balance.

E.H. Smith Contracting Co. v. New York, 240 N.Y. 491 (1925), involved the construction of an elevated railway. The contract provided that the engineer would, in all cases, determine the amount, quality, acceptability, and fitness of the work and materials. The engineer would also determine all questions with regard to the work and the construction thereof. His determination and estimates would be final and conclusive upon Smith. The court found that the engineer's action involved an erroneous construction of the contract. Where there is no reasonable basis for an engineer's action, and it is patently erroneous, then the courts find the equivalent of bad faith on the part of the engineer. The contractor cannot be bound by his decision. The court found for Smith.

The courts will consider all the facts in determining the reasonableness of an architect's decision. In *Fuchs v. Saladino*, 133 A.D. 710, 118 N.Y.S. 172 (1st Dept. 1909), a contract balance was sought from Saladino in connection with the construction of a seven-story building. There were various defects and deviations from the contract drawings and specifications. The court found that approximately 15 percent of the value of the work remained unperformed. Under those circumstances, it held that the failure of an architect to certify that the contractor has performed his or her contract cannot be deemed arbitrary or unreasonable. The architect's certificate is a condition precedent to the contractor's right to recovery. Where the contract is neither completely nor substantially performed, the contractor cannot argue that he or she is excused from procuring the architect's certificate.

Who makes the decision as to whether the architect's or engineer's refusal to issue a certificate is reasonable? The jury, of course. What is the criterion which the jury employs in making this determination? *Becker v. Woarms*, 72 A.D. 196, 76 N.Y.S. 438 (1st Dept. 1902), appeal withdrawn 179 N.Y. 515 (1904), involved a dispute with respect to certain bronze and metal fittings to be manufactured and installed in a bank building. The contract contained a provision that any questions regarding the work should be referred to the architect. The architect's decision would be binding. In addition, prior to final payment, the contractor was to obtain a written certificate from the architect. The question before the jury was not whether after reviewing the labor performed and materials furnished they would have decided as the architect decided, but rather whether the architect, in good faith, denied the contractor's request for a certificate of completion. The issue is not reasonable compliance with the plans and specifications, but rather whether the architect's refusal is reasonable.

§ 3.5.5 Delegation of Approval

Where the contracting parties agree that certain matters are to be finally determined by an engineer or architect, what are the responsibilities of that engineer or architect in making this determination?

In *Sweet v. Morrison*, 116 N.Y. 19 (1889), reh'g denied 23 N.E. 897, reh'g denied 24 N.E. 1097, discussed at § 3.5.2, the responsibilities of the chief engineer were discussed by the court. As you recall, *Sweet* involved the construction of some 200 miles of road and track for the Northern Pacific Railroad Company. The contract provided that the chief engineer would make the final determination on certain matters. He was to be the sole judge of the quantity of work done and the materials furnished. He was not to measure, but rather to estimate. The court found that the parties evidently meant that he was to act in a possible and practicable way. How was he to make his estimate if not from personal measurement and observation? The obvious answer is that it was necessary that he rely on the reports of his subordinates. There was no other way of estimating the amount of work extending over miles of territory. Even if one person could do it, the court observes that the head of the engineering department that was building a railroad across the continent would not be selected for that purpose. As chief engineer, he was conversant with the general facts and had a thorough knowledge of the engineers under his supervision. He was in a position to exercise judgment on the reliability of their reports and could direct others to revise their measurements if he deemed it appropriate. The evidence showed the estimates were made from actual surveys and measurements by engineers of the railroad company in the presence of the contractor's foreman.

Where the contract provides that certain determinations are to be made by the chief engineer, it is not necessary that he personally make the estimates and measurements upon which those determinations are based.

Does the death of the architect allow the contractor to avoid the requirement of a certificate?

Beecher v. Schuback, 4 Misc. 54, 23 N.Y.S. 604 (N.Y. Co. 1893), aff'd 1 A.D. 359, 37 N.Y.S. 325 (1st Dept. 1896), aff'd 158 N.Y. 687 (1899), involved a building construction contract. When the architect who was to provide a certificate for final payment died, another architect was substituted by the owner. The contractors accepted this architect. Clearly, his certificate would stand for that of his predecessor. The death of the original architect did not excuse the need for a certificate for final payment.

§ 3.5.6 Condition Precedent

As discussed at § 2.15.2, the courts will uphold contractual provisions that establish a condition precedent to payment. Will the court uphold other conditions? The answer is generally yes.

In *Gallagher v. Baird,* 54 A.D. 398, 66 N.Y.S. 759 (1st Dept. 1900), aff'd 170 N.Y. 566 (1902), Gallagher had contracted to construct an armory. Baird and Sargent Granite Company had executed a bond for the benefit of Gallagher that guaranteed that Sargent would fully perform a subcontract with Gallagher to provide granite for the erection of the armory. The granite was to be in conformance with the specifications of the contract. The subcontract further provided that before any payment would become due, Sargent would provide to Gallagher a certificate of the supervising architect approving the quantity and quality of the granite. Neither Sargent nor Baird ever provided a certificate of the architect showing his approval. The court found where the materialman failed to procure the architect's certificate, he could not compel the contractor to make a payment. The obtaining of the architect's certificate was a condition precedent to payment.

In *Arena Constr. Co., Inc. v. Town of Harrison,* 71 A.D.2d 647, 419 N.Y.S.2d 3 (2nd Dept. 1979), Arena contracted to construct a townhouse. Certain claims were to be submitted by Arena to the architect for resolution. Under the terms of the contract, the architect's determination was final and conclusive. However, recall that the courts make an exception if that determination is made through fraud, bad faith, or failure to exercise honest judgment. Arena alleged that the rejection of its claims by the architect was arbitrary and not in accordance with the facts. Arena is entitled to the opportunity to prove that the architect's determination was, in fact, arbitrary. The court noted that if Arena fails to prove that the architect's decision was arbitrary, then that decision is final and conclusive.

Bistrian Gravel Corp. v. Wainscott Northwest Associates, 116 A.D.2d 681, 497 N.Y.S.2d 748 (2nd Dept. 1986), involved a contract for the insulation of three fire cisterns. Bistrian installed three concrete cisterns. The fire department having jurisdiction, rejected the cisterns on the basis that they were concrete and not fiberglass. The oral agreement between Bistrian and Wainscott did not make the approval of the fire department a condition precedent for payment. The court will not imply a condition precedent. The court determined that Bistrian was entitled to be paid for installation of the cisterns.

§ 3.5.7 Grounds for Attack

A contract may make an architect's or engineer's certificate a condition for payment. The contract may also provide that the architect's or engineer's decision on certain aspects of the contract is final and conclusive. How does a contractor attack such a decision?

We have seen that the cases provide that the decision must be free from fraud, that it must be made honestly, and without such mistakes as to imply bad faith. Where the courts find bad faith, the contractor is not bound by the architect's or engineer's decision. Where the engineer has made an erroneous assumption in interpreting the contract, the contractor is not bound by that interpretation. *Savin Bros., Inc. v. State*, 62 A.D.2d 511, 405 N.Y.S.2d 516 (4th Dept. 1978), aff'd 47 N.Y.2d 934, 419 N.Y.S. 2d 969 (1979). Additionally, where it is alleged that the certificate was procured through fraud, evidence of such fraud must be allowed to impeach the certificate. See *Byron v. Bell*, 10 N.Y.S. 693 (N.Y. Co. 1890).

§ 3.5.8 Owner Preventing Issuance of Certificate

Naturally, where the owner prevents the issuance of a certificate, he cannot then demand that certificate as a condition to payment. The owner's interference excuses the contractor from producing the certificate.

In *Pneumatic Signal Co. v. Texas & Pacific R. Co.*, 200 N.Y. 125 (1910), Pneumatic had undertaken to construct and install a system of electric interlocking railroad signals. The contract required that payment would not be made unless the signal system was accepted by the Railroad Commission of Texas. Although Pneumatic did not prove such acceptance, it was improper to allow the railroad to claim such certificate was essential for recovery. Pneumatic showed that the failure of the railroad commission to approve the signal system was due to the omission of the railroad to make certain additions and improvements. These were required by the railroad commission and were the duty of the railroad, not Pneumatic, to make. The failure of the railroad commission to finally approve the system was not due to any default, negligence, or omissions on the part of Pneumatic. Pneumatic did not fail to comply with the provisions of the contract. The evidence showed that the failure of the railroad commission to accept the signal system was wholly due to the default, negligence, and omission of the railroad.

§ 3.5.9 Waiver of Certificate

A contractual provision may, of course, be waived by the conduct of the parties. A requirement that an acceptance by the architect is a condition precedent to any payment can be waived. *A.D. Granger Co. v. Brown-Ketcham Iron Works*, 204 N.Y. 218 (1912).

Does the requirement that an architect's certificate be obtained apply in the situation where a contractor properly ceases work on a contract and seeks recovery for the work performed?

In *Borup v. Von Kokeritz*, 162 A.D. 349, 147 N.Y.S. 832 (2nd Dept. 1914), later appeal 184 A.D. 957, 171 N.Y.S. 1079 (1918), Borup had begun to build a house for Von Kokeritz. The general rule is that if performance of the contract has been stopped by the fault of the owner, then the contractor can elect to rescind and sue for the reasonable value of his work or, in the alternative, to sue for the work done according to the contract and for the losses of profits, which he has suffered.

If the cost to complete the work and the amount paid to the contractor is less than the contract price, then the owner will be liable to the contractor for the contract price, less prior payments to the contractor, less the amount the owner paid to complete the contract. The amount remaining on the contract is deemed to be held in trust for the contractor. See *White v. Livingston*, 69 A.D. 361, 75 N.Y.S. 466 (2nd Dept. 1902), aff'd 174 N.Y. 538 (1903).

Where the owner has improperly stopped the work, the contract requirement for a certificate of the architect has no application. Borup was entitled to a verdict for the difference between the original contract price, less prior payments and the cost to the owner to complete the contract.

Hoisting Machinery Co. v. Federal Terra Cotta Co. 179 A.D. 653, 167 N.Y.S. 85 (1st Dept. 1917), aff'd 228 N.Y. 569 (1920), involved a contract under which Hoisting Machinery agreed to deliver to Federal certain materials, which were to be used in construction of an overhead monorail trolley system in Federal's plant in Woodbridge, New Jersey. The contract provided that approval was to be given by Federal's engineer for the work carried out under the contract specifications. Federal's defense to payment was that Hoisting Machinery never procured the approval of Federal's engineer and, therefore, was not entitled to final payment. The evidence was that material provided was not defective. The materials were furnished, accepted, and used by Federal. Federal's engineer failed to express disapproval thereof. Where the workmanship and the materials are substantially in conformance with the contract, the court finds that it is not material whether the engineer did or did not expressly approve the workmanship and materials. Hoisting Machinery prevailed.

Where the contract requires an architect's certificate as a condition for payment, may the contractor recover where the owner took the work out of his or her hands and completed it? The answer is yes. In *Dwyer v. Mayor*, 77 A.D. 224, 79 N.Y.S. 17 (1st Dept. 1902), Dwyer was engaged to construct a boiler house and engine room for the Metropolitan Museum of Art. The contract was performed in complete conformance to the plans and specifications. The architect withheld his certificate, insisting that the work had not been properly performed. The dispute centered about the construction of a boiler flue.

Dwyer constructed the flue in accordance with the contract plans and specifications. Early in the construction, he notified the architect that the flue, if constructed to the plans and specifications, would not be watertight. He was subsequently directed by the architect to complete the flue without any changes or amendments to the plans and specifications. After completion, it was ascertained that the flue was not watertight. The architect then materially altered the plans and specifications and directed Dwyer to take the flue down, remove it and reconstruct it. Dwyer, insisted that he had performed his contract and was under no obligation to perform the work again in a different manner. Dwyer appealed to the City Board of Park Commissioners. His protest was to no avail. Dwyer was directed to reconstruct the work as required by the architect. Dwyer then notified the architect and the Board of Park Commissioners that the flue would still not be watertight, if constructed according to the amended plans and specifications. The reconstructed flue was not watertight.

The contract had expressly provided that the work would be done to the satisfaction of the Commissioners and the architect. The specifications stated that the flue was to be impervious to water and watertight. The legal effect of this contract was that Dwyer undertook to construct the flue in accordance with the plans and specifications. Dwyer was to make it watertight so far as construction in accordance with those plans and specifications would produce a watertight flue. Dwyer did not guarantee the efficacy of the plans and specifications in this regard.

The court found that Dwyer had performed his contract with reference to the construction of the flue and that it was the architect's blunder in preparing plans and specifications that prevented the construction of the desired watertight flue and that Dwyer made out his case to cover the cost of reconstructing the flue.

3.6 WAIVER

§ 3.6.1 Definition

When the issue of waiver is brought before a court, the court will examine the waiver to determine whether or not it was intentional. Waiver is a matter of intent. In order to waive a right or advantage the person possessing that right or advantage must intend to waive it. Negligence, oversight, or thoughtlessness will not create a waiver.

In *Alsens American Portland Cement Works v. Degnon Contracting Co.*, 222 N.Y. 34 (1917), the facts were undisputed. This is often the case where the issue of waiver is raised. If the parties to the contract agree on the facts, the issue of waiver will be raised. The question, then, is whether an evaluation of the facts indicates presence of intent: that necessary element of waiver.

Pursuant to their contract, Degnon was to purchase from Alsens 175,000 barrels of cement to be totally ordered for shipment by a certain date at a certain price. By the date certain, there remained unordered 36,000 barrels. Subsequently, Degnon ordered and received from Alsens 1,900 barrels. Alsens seeks the increased market value for those 1,900 barrels, except for the first two shipments. Degnon conceded that it was bound to order the concrete before the date certain. After that date, Alsens could refuse to make any delivery under the contract. Degnon's argument was that Alsens waived its right to refusal by delivering the cement.

The court noted that a waiver is an intentional abandonment or relinquishment of a known right or advantage. The party would have enjoyed that right or advantage, but for the waiver. The intent to relinquish the right or advantage must be proved.

The court notes that waiver can be established in two fashions. It may be established as a matter of law where there is an expressed declaration of the person against whom waiver is sought, or that person's undisputed acts or language is so inconsistent with his or her standing upon his or her rights, as to leave no opportunity for a reasonable inference to the contrary. On the other hand, and more commonly, waiver is proved by various declarations, by acts, or by nonfeasance, all of which permit differing inferences that do not directly, unmistakably, or unequivocally establish it. It is then for a jury to decide whether or not there was waiver as a matter of fact. The court in *Alsens*, held that the jury must decide whether the facts established waiver. The court found no express waiver and determined that waiver was a question of fact.

Arrow Plumbing Co. v. Dare Constr. Corp., 212 N.Y.S. 2nd 438 (Nassau Co. 1961), involved a dispute between a plumber and the general contractor. The subcontract contained a clause inserted for the protection of Arrow, the plumbing contractor that stated, "No additional work will be done outside this contract unless it is in writing and signed for before it is done." The court found that this clause was inserted to excuse the nonperformance of the plumbing contractor in the absence of a written order. A party to building contract can waive a provision in that contract inserted for his benefit.

The court notes that even if the clause is construed to have been inserted for the protection of the owner, a representative of the owner was present each time the additional work was discussed. Therefore, the condition may be deemed to have been waived. The court found that the extra items not covered by the contract, and sought by Arrow, were explicitly ordered by the general contractor with the knowledge and consent of the owner. The court found for Arrow Plumbing.

Hevenor v. Union R. Co., 204 A.D. 535, 198 N.Y.S. 409 (1st Dept. 1923), involved a dispute over a surety bond required by a contract. Hevenor sought its balance on a contract for the reconstruction of street railroad tracks. A provision of the contract had required that Hevenor provide a bond to indemnify and hold harmless the railroad company. The bond was not given and Hevenor claims that the provision was waived.

Prior to beginning the work, Hevenor wrote to the railroad and asked that the provision be waived. The railroad said it would not waive the bond and requested that Hevenor furnish the same, or a certified check in the bond amount. Neither was furnished by Hevenor.

The contract stated that, "The contractor herewith delivers...a bond... ." The court points out that this is an acknowledgement of the receipt of the bond. While the railroad company could have disclaimed liability on the contract on a showing that it did not receive the bond prior to the beginning of the work; it could not go forward with the work, accept its benefits, and after completion assert it had not receive the bond. The railroad's conduct indicated that it never intended to receive the bond.

The waiver is proved as a matter of law by undisputed acts or language of the party against whom the waiver is asserted. The courts holds that waiver is established as a matter of law. See also *Arndt v. Leff*, 14 Misc.2d 677, 179 N.Y.S.2d 356 (Albany C. Ct. 1958).

§ 3.6.2 Retention and Use

Can an owner claim that you have not properly performed when he or she takes possession of the object of a contract and retains it for his or her own use? Generally, an owner who takes possession of manufactured goods does not waive his right to claim they were not manufactured properly. There is a caveat, however. *Vandegrift v. Cowles Engineering Co.*,161 N.Y. 435 (1900), involved a contract for the construction of a steamboat. While the shipbuilder was still working on the steamboat and before the time for the completion of the contract had expired; possession of the vessel was taken, it was removed to another state, and completed. The court noted that the shipbuilder could have regarded this as a trespass. In the alternative, he could regard the taking as an acceptance under the provisions of the contract. The vessel was taken before the period in which there was a right to reject it and there was no proof of any rejection. The court found that taking possession of the vessel could only be rightfully accomplished if it was accepted under the contract. Having thus elected to accept the boat, the contrary can no longer be asserted. A liability cannot be imposed that might not have otherwise arisen.

§ 3.6.3 Use of Building

The seminal case in this area is *Cawley v. Weiner*, 236 N.Y. 357 (1923), which involved the construction of a bungalow. Cawley sought his contract balance and certain extras from the Weiners. The Weiners entered into possession of the bungalow and prepared a punch list of items they considered necessary for completion of the contract. At trial the Weiners also attempted to prove that they subsequently discovered that the floor of their bungalow sagged, that there was no supporting arch under the living room fireplace, and that there was a half-inch crack in the brick wall of the first floor. The trial court had found that the Weiners had waived these additional three objections. The highest court in New York disagreed, indicating that this was not a sale of goods. The house was built upon the defendants' property. The Weiners could move into it, live in, and accept it without waiving any defects in construction. Citing a Massachusetts case, *Gray v. James*, 128 Mass. 110, the court found that the Weiners did not waive the various defects and reversed the award to Cawley. See also, *Meyer v. Woodward-Brown Realty Co.*, 209 A.D. 548, 205 N.Y.S. 114 (1st Dept. 1924), aff'd 239 N.Y. 613 (1925).

An important twist is contained in *Beagelman v. Lydia Homes, Inc.*, 11 Misc.2d 158, 171 N.Y.S.2d 279 (Nassau Co. 1958). Beagelman sought damages, alleging that Lydia Homes failed to construct a dwelling house and to grade the premises according to contract. However, the contract provided that none of the terms, unless expressly indicated, would survive delivery and acceptance of the deed.

All conditions complained of were known of by Beagelman, who had lived in the house prior to accepting title. Before transfer of title, Beagelman's attorney addressed a letter to Lydia Homes enumerating the defects and stating that they must be corrected and adjusted prior to any final delivery of title.

Some of those items in the letter are those about which Beagelman now complained. Beagelman, however, did accept title without any written reservation. The court found that since the defects were not latent, and since Beagelman had accepted the provisions of the contract, the acceptance of title is a waiver of all objections.

§ 3.6.4 Acceptance

Mohawk Overall Co. v. Brown, 163 A.D. 157, 148 N.Y.S. 369 (3rd Dept. 1914), involved the construction of a factory building for Mohawk Overall. The building was to be approximately 36 feet by 137 feet, and four stories in height. The contract plans and specifications detailed the construction of a modern manufacturing plant with heavy walls, concrete floors, and steel girders.

The building was a failure and began showing signs of weakness almost immediately. In less than two months, it became necessary to reinforce the building in order to make it safe to occupy. Mohawk Overall complained that the workmanship was not up to the standard required by the contract and that the cement used in laying the brick walls did not meet the specifications. Brown, the builder, claimed that the bricks specified by the contract were not the quality that should have been used, and that the architect, under whose supervision the work was being done, knew the class of workmanship being put in and the mortar being used. Brown urged that if the building constructed under the supervision of Mohawk Overall's architect was not what was required, then the fault was in the plans and the approval of the architect.

The court found that there was no question that the workmanship was not up to the standards of the contract. The construction of this factory was so poor as to evidence bad faith, with a reckless disregard of the requirements of a building in which "...human beings are called upon to earn their daily bread... ." Using a public policy argument, the court found that Brown should not be permitted to sustain his defenses, even though the architect was empowered to waive the conditions of the written contract. The court finds that no one could successfully maintain that the work of the character expended could have been furnished in good faith. Since the poor workmanship was responsible for a dangerous and unsatisfactory condition, the court awards the cost of correction to Mohawk Overall.

§ 3.6.5 Waiver

As we have previously discussed, a party who benefits from a provision of the contract may waive that provision. However, once that party waives that provision, he or she cannot rely on it to prevent recovery where he or she has caused or consented to nonperformance.

M. O'Neil Supply Co. v. Petroleum Heat & Power Co., 280 N.Y. 50 (1939), involved a dispute between Petroleum, heating pipe installation contractor, and O'Neil a material manufacturer. Petroleum was installing a heating plant in a hotel. Petroleum's subcontractor, Charles, was to do certain pipe installation and other work. Charles had purchased certain materials from O'Neil and owed a balance. O'Neil was about to file a lien against the hotel and against Petroleum and Charles. Charles had stopped work because of disagreements with Petroleum and had filed a lien against the hotel.

O'Neil, Charles, and Petroleum entered into a three-way agreement. Pursuant to the agreement, Charles agreed to waive any right on the lien filed and to any claim against Petroleum for past or future labor and services upon payment to Charles. Charles agreed to promptly proceed with and complete the work remaining to be done. O'Neil agreed to furnish the necessary materials for completion of the work and to waive any right to lien for any further claim for materials already furnished or to be furnished. In return, O'Neil was to be paid by Petroleum, upon request, five days after completion of the installation by Charles. Although the necessary materials were furnished, Charles failed to complete the installation. Petroleum refused to pay the amount due to O'Neil, solely on the grounds that Charles did not complete the work. Petroleum claimed that the completion of the installation by Charles was a condition precedent to its liability to payment.

The court noted that O'Neil had no pecuniary interest in the completion of the work by Charles. The failure of Charles was at no fault of O'Neil. Payment by Petroleum to O'Neil could not definitely be withheld because of nonperformance by Charles. The court noted that every contract implies a good faith and fair dealing between the parties. The court finds that Petroleum cannot rely on a condition precedent to prevent recovery where the nonperformance of that condition was caused or consented to by itself. Only Petroleum could compel Charles to perform. It was Petroleum's duty to see that the work was completed within a reasonable time. Payment to O'Neil cannot be indefinitely withheld.

In *Mogulewsky v. Rohrig*, 104 A.D. 147, 93 N.Y.S. 590 (2nd Dept. 1905), the last payment for construction of a building was to be made when a permanent loan was obtained for the building. The owner did not obtain such a loan. The court found that this defense to payment was not available to the owner because he was solely at fault for not having obtained the loan. The contractor was to receive the balance for his work, labor, and services performed, and materials furnished in construction of the building.

§ 3.6.6 Waiver of Time of Performance

Can contract provisions regarding the time of performance be waived? They can be waived both expressly or impliedly. How is the time provision of a contract waived? There are a number of ways. Waiver may result when a contract begins after its time limit has expired. In addition, contract performance may continue after the time limit has expired. Acceptance of the contract after the time limit has expired will also imply a waiver.

General Supply & Constr. Co. v. Goelet, 241 N.Y. 28 (1925), remittitur am'd 241 N.Y. 507 (1925), involved a contract for masonry, carpentry, structural iron, and steel work. The contract included a completion date and a liquidated damages clause. The work was not completed on the completion date and General Supply was allowed by Goelet to continue performance of the work. In this, Goelet waived time as an essential element of the contract. Having done so, Goelet could not then suddenly abandon that position and treat as essential an element of the contract, which he had previously waived. Although Goelet could not use failure to meet the completion as a grounds for termination, he could have restored time as an essential element of the contract by a notice to General Supply to complete within a reasonable time.

The work was far from finished when the owner, in spite of General Supply's protest and resistance, ejected General Supply from the premises and prevented it from proceeding with the work. This the owner could not do. The owner should have proceeded pursuant to the contract. The contract required that the work could be terminated at any time upon certificate of the architect that the work was being unreasonably delayed and that such delay was sufficient grounds for termination.

As a contractor, the procedural issue in this case should be born in mind. Most subcontracts provide for three-day notice for termination. The termination provision should be closely followed, should termination prove unavoidable.

An explicit waiver is found in *Emslie v. Livingston*, 51 A.D. 628, 64 N.Y.S. 259 (2nd Dept. 1900). Emslie had agreed to complete a dwelling house and stable for Livingston. Livingston resisted payment for labor and materials, claiming a failure by Emslie to complete the work in the period required by the contract. At some point during the construction, payment had been refused to Emslie, the site was abandoned, and materials withdrawn. In ensuing negotiations, a settlement was reached. That settlement must be regarded as final and conclusive with respect to prior claims and transactions. The settlement was a waiver on the part of Livingston of the time limit and an abandonment of all claims for delay damages.

Where a delay is caused by a breach of contract by the owner, a construction contractor will not waive his or her right to recover damages because he or she proceeds with work under the contract. This is especially true where the contractor has protested the delay.

Belmar Contracting Co. v. State, 194 A.D. 69, 185 N.Y.S. 734 (3rd Dept. 1920), rev'd on other grounds 233 N.Y. 189 (1922), concerned a contract for the improvement of 1.17 miles of highway. Belmar executed the contract and filed the required bond. Although these documents appeared satisfactory to the Commissioner of Highways, the Commissioner delayed the execution of the contract on the part of the

State. It was impossible to complete the work before winter. By mutual consent, the work was postponed until the following spring. The work was commenced and completed, but Belmar had given notice of the delay and sought the increased costs.

An express waiver causes the contractor's downfall in *Cascade Automatic Sprinkler Corp. v. State*, 147 Misc. 420, 263 N.Y.S. 805 (Ct. Cl. 1933), aff'd 247 A.D. 226, 286 N.Y.S. 698 (3rd Dept. 1936), modified on reh'g 248 A.D. 922, 290 N.Y.S. 382 (3rd Dept. 1936), aff'd 277 N.Y. 612 (1938). Cascade's contract with the State was for the installation of a heating system and incidental work on two hospital buildings. Upon completion of the work, the State's chief engineer certified that Cascade was entitled to a final payment. An order was made by the Secretary of the Department of Mental Hygiene approving said payment. Cascade signed a receipt for said payment.

Cascade's contract with the State contained an article providing that acceptance of final payment would constitute a waiver of all claims by the State and of all claims by the contractor. The court found that the execution by Cascade of the receipt for the final payment, which was marked "in full," was an express waiver of any claim for damages it might have had.

The court found that there was no legal obligation upon the State to pay. The court further noted that there was no moral obligation resting upon the State to pay extra compensation. Such a payment would be a gratuity, which is forbidden by the Constitution.

§ 3.6.7 Effect

An important distinction must be borne in mind. By permitting a contractor to proceed with work after the completion date, the owner does not lose his or her right to recover damages for the delay, but at most, loses his or her right to terminate the contract. Of course the owner can correct this situation by providing the contractor with written notice to complete within a reasonable time.

Kelly v. St. Michael's Roman Catholic Church, 148 A.D. 767, 133 N.Y.S. 328 (2nd Dept. 1912), involved the construction of church and rectory buildings. Kelly sought to be paid for the contract balance. St Michael's resisted payment, claiming that Kelly had not completed the work by the completion date. The court held that where there is a delay in performance of the contract, and the owner does not insist upon his strict legal right to put an end to the contract for failure to complete by the completion date, and instead permits the contractor to continue his work, then the owner will not be allowed to interpose the contractor's delay as a defense to payment.

§ 3.6.8 Modification

The parties to a contract may modify that contract, thereby waiving certain of the contractual provisions. Although an owner may waive strict application of the contract provisions, he or she is not necessarily entitled to be compensated for that waiver.

In *Thomas W. Finucane Co. v. Board of Education*, 190 N.Y. 76 (1906), Finucane had contracted to erect a high school. The specifications required that all finished floors be kiln dried, taken directly from the kiln to the machine for planing and matching, and immediately taken into the school building for laying. These specifications, as they related to the place where the flooring was to be kiln dried, were waived by the Board of Education.

Subsequently, the Board of Education sought to be reimbursed for the contract savings. The court noted that the Board's waiver of the contract provision that the flooring should be kiln dried on the construction site was unconditional. Although it was clear that the intention of Finucane in requesting a waiver of this provision was doubtless for the express purpose of enabling him a better bargain in purchasing lumber kiln dried at the source, in no event was this a basis for allowing reimbursement to the Board.

The Board did not claim that the flooring absorbed any moisture in being transfered to the school building, nor that it was in any way injured or inferior to the floors required by the specifications. The flooring was as contemplated by the specifications and, in every respect, according to the specifications. The waiver of the kiln drying provision was unconditional. There was no basis to reimburse the Board.

In *Logan v. Consolidated Gas Co.*, 107 A.D. 384, 95 N.Y.S. 163 (2nd Dept. 1905), Logan had contracted to provide the Gas Company with a quadruple-section gas holder with steel tank. During a testing period the tank collapsed. Logan argued that bills of sale were delivered to the Gas Company and that, by the company's acceptance of those bills of sale, the contract was modified. The court held that Logan contracted to deliver a completed structure and can only recover by showing performance and acceptance of the completed structure, or waiver by the Gas Company. The delivery of bills of sale as the work progressed does not alter the situation. It does not amount to a modification of the contract, or an acceptance of the work done, absent proof of an act evidencing an intention to accept the work, or waive or modify any provisions of the contract.

§ 3.6.9 "No Damages" Clause

As extensively discussed at § 2.15.1, you are not entitled to reimbursement for delay if the cause of that delay is within the scope of a no damage provision. *Taylor-Fichter Steel Const. Co. v. Niagara Frontier Bridge Com.*, 261 A.D. 288, 25 N.Y.S.2d 437 (1st Dept. 1941), aff'd 287 N.Y. 669 (1941), involved a contract to erect the superstructure of two bridges being built over the Niagara River near Buffalo, New York. Another contractor was to erect the supporting substructures consisting of various abutments and piers.

Taylor-Fichter claimed it could have completed its work within 14 weeks if permitted to proceed continuously. Instead, the work was spread over 37 weeks. There were explicit contractual provisions that the Bridge Commission would not be liable due to delay on the part of the substructure contractor. Taylor-Fichter's damages included items for prolonged retention of machinery and equipment, and additional cost for labor for overtime.

The contract specifications did contain a schedule showing the dates on which the substructure contractor was to complete various abutments and piers. However, the specifications also provided that Taylor-Fichter was to have no claim against the Bridge Commission other than for an extension of time because of the failure of the substructure contractor to have an abutment or pier finished on the dates mentioned. The cause of this delay was so clearly contemplated by the parties that the court found that Taylor-Fichter was not entitled to its damages.

The discussion of no damages for delay provisions at § 2.15.1 demonstrates the courts' clear bias against the contractor. In *Novak & Co. v. New York City Housing Authority*, 108 A.D.2nd 612, 485 N.Y.S.2d 68 (1st Dept. 1985), appeal dismissed 67 N.Y.2d 1027, 503 N.Y.S.2d 326 (1986), Novak was the plumbing contractor at a public housing construction project. Novak sought reimbursement for increased labor costs, increased superintendent costs, and increased field and home office costs because of delays. Novak contended that the delays were a result of the City's failure to properly coordinate work with the general contractor, failure to provide elevator service, its disruption of the normal sequence of operation, its failure to timely process and prosecute its part of the work to be performed, misrepresentations, and its failure to accept work in accordance with the contract. The court found that these causes were excluded by the no damage provision, unless the City's conduct was motivated by bad faith, malicious intent, or amounts to gross negligence. Accordingly, the court found that the contractor's damages were barred.

§ 3.6.10 Payment

Can the fact of final payment constitute waiver of a contractual provision? The mere fact of payment will not prevent an owner from recovering for defects subsequently discovered. *Asbestolith Mfg. Co. v. Kerley*, 129 N.Y.S. 512 (App. T. 1911), involved the installation of flooring at a summer home. Asbestolith sought its contract balance. Kerley, arguing that the floors were not hard and firm, became soft and discolored, and retained the marks of shoe prints. The floors were rendered practically useless. After the laying of the floors was completed, the contractor was paid $1,000.00 on account. The court specifically found no acceptance of the work by the owner. Even if payment on account were to be deemed an acceptance, it would not, under the circumstances of the case, be deemed in law an absolute acceptance. At the time the payment on account was made, the durability of the work, permanency of the coloring of the floors, their hardness and firmness when subject to ordinary use, could not be determined. The court held against the contractor.

Diehl v. Schmalacker, 30 Misc. 786, 62 N.Y.S. 1080 (App. T. 1900), involved a contract for masonry work at three buildings. The contract required that the work was to be done to the plans and specifications and to the satisfaction and under the direction of the owner's architect. That satisfaction was to be testified to by a writing or certificate. The court noted that Schmalacker had waived various departures from the contract. Schmalacker saw the work in progress, noticed departures during construction, yet made voluntary payments. By those payments, he is deemed to have waived the defects.

However, payment by the owner does not always mean a waiver of his or her contract. *Witt v. Gilmour*, 172 A.D. 110, 158 N.Y.S. 41 (2nd Dept. 1916), involved the construction of a country home and garage. Although periodic payments were made upon certificates of the owner's architect, the owner did not waive defective construction. The owner was absent most of the time. In addition, the contractor stated that no other payments were due until completion and that he would remedy any defects and deduct their cost from the final payment.

There were substantial defects and deviations. The construction of the chimney was extremely loose, and in certain areas only a flue liner covered with stucco was installed. In addition, timber headers penetrated dangerously close to the chimney flues. The flue linings, so important because of the slack masonry surrounding them, were set irregularly to leave cracks between the sections through which sparks might escape.

The foundations were insufficient. Footing courses were specified to be of brickwork under chimneys, brick walls, and piers for solid girders. Instead, they were of concrete, improperly laid, contained rubbish, brickbats, and cement so poor that it crumbled in one's hand. Entire walls contained no reinforcing rods or concrete. Heating and plumbing work was not completed. Iron pipe and window glass were of an inferior quality. Instead of conforming to specifications for a cement, gravel, and sand mixture to deafen the bathroom floors, the mason used a mixture of cinders, which tended to destroy water pipes by corrosion.

Witt was the architect for the house, but subsequently undertook as contractor to build it. Witt's contention was that, owing to the certificates he issued, there was a waiver of Gilmour of these defects. This could not be sustained. The first certificate was issued before work on the chimney had begun. The Gilmours were absent during much of the following intervals. After their return, the Gilmours visited the house only once before payment on the second certificate was made. At that time, Witt had said that the contract did not make payments final until all payments were made. In addition, Witt stated that he would correct any mistake or defect. Under these circumstances, the payments will not effect a waiver of the defects.

3.7 IMPOSSIBILITY OF PERFORMANCE

§ 3.7.1 Contracts with Alternatives

Subsequent impossibility will generally not allow nonperformance of a contract. There is an exception if the impossibility is provided for in the contract. However, courts throughout the nation have sanctioned various exceptions to the rule. Destruction of a building or of the property subject to a contract, the termination of existence of the contracted for object, inevitable accidents, strikes, insolvency, death, sickness, blockade, and war have all been considered by the courts. Whether the court find the performance excused or not will often hinge on the particular facts and the particular jurisdiction. This is an area where the expertise of your local attorney is most important.

Buffalo & L. Land Co. v. Bellevue Land & Improv. Co., 165 N.Y. 247 (1901), involved an agreement by Bellevue to build, maintain, and operate an electric street railroad. Buffalo complained that Bellevue failed to operate the railroad during a particularly severe winter storm. The court found that the railroad had become blocked with snow and snowdrifts caused by heavy snowfalls and high winds. It was practically impossible to run the cars on the railroad while the snow and snowdrifts continued. However, once the snow was removed, the regular running of the cars was resumed and continued until the drifts once again blocked the tracks.

The court notes that it is a well-settled rule of law that when a person engages to do an act and does not provide against contingencies in the contract, he or she will not be excused by an inevitable accident or other contingency. But, the court notes that there are many contracts which, by their very nature, imply that the contractor will be relieved from consequences of nonperformance where such performance is rendered impossible without that person's fault. The court finds that the contract in question belongs in that category.

The agreement to run the cars was qualified, not only by the nature of the contract, but by the words, "as such street railroads are usually run." The court notes that the operation of railroads is frequently interrupted by storms and, therefore, finds that the railroad's performance was adequate in view of the conditions it encountered.

As a general rule, impossibility will not excuse nonperformance of a contract. Although a number of exceptions to this rule have developed, the category of strikes is not among them. In *Kinser Const. Co. v. State*, 125 N.Y.S. 46 (Ct. Cl. 1910), aff'd 145 A.D. 41, 129 N.Y.S. 567 (3rd Dept. 1911), aff'd 204 N.Y. 381 (1912), which involved the construction of a portion of the New York State Barge Canal System, the court noted all exceptions that would excuse nonperformance of the contract. Performance is excused where legal impossibility arises from a change in the law, or where the specific thing essential to performance is destroyed, or where there is an incapacity by sickness or death in the case of a contract for personal services, or where a contingency arises that is not covered by the express terms of the contract and that renders the performance of the contract as planned impossible.

While Kinser was excavating for one of the canal locks, an extensive cave-in occurred. The earth was of a slippery, greasy clay with insufficient resistance to permit construction according to the contract plans and specifications. Kinser's expert testified that he had never seen soil like it and that it would not be good engineering practice to build a lock in such material. The court felt that these circumstances called into play the fourth exception. Here, in the course of construction of a canal, natural conditions of soil unexpectedly appeared, which the contingencies stated in the contract did not expressly cover. Performance of the contract as planned was rendered impossible. The necessary and significant changes in the nature and cost of the contract substantially affect the work remaining under the contract. The law will read into the contract an implied condition, that such a contingency will terminate the entire contract.

Conditions that render performance impossible will not terminate the entire contract, nor will they vitiate what has been done. The extent of the conditions may amount to a substantial abrogation of the entire contract or may relate to an insignificant part of the contract. They excuse performance only to the extent to which performance is impossible. They leave what has been done valid and permit recovery for what has been done. Performance is excused only to the extent to which performance is impossible. In *Kinser*, the State was not in a position to compel performance of an impossibility. Kinser was entitled to recover for the work he had performed.

In *Monroe Piping & Sheet Metal, Inc. v. Edward Joy Co.*, 138 A.D.2d 941, 527 N.Y.S.2d 279 (4th Dept. 1988), Monroe, the subcontractor, conceded that it could not perform the subcontract work. Monroe argued that the subcontract was impossible to perform because of a strike against Edward Joy Company by the steamfitters union. Realizing that the strike was affecting Monroe's employees, Edward Joy had provided a separate gate for Monroe employees to enter the construction site. In this fashion they would not cross the picket line. The court found that, at best, Monroe could assert that performance of the subcontract was difficult because of the strike. Monroe had not shown that performance was impossible.

Kinser demonstrated that impossibility only excuses the portion of the performance that is impossible to perform. Suppose the contract contains alternative performance? For example, the contractor is to install either a gas operated or an oil operated furnace. The choice is the contractor's. However, if it is impossible to install a gas operated furnace, all performance under the contract is not excused. The contractor must still perform the portion that is not impossible. He must install the oil operated furnace.

§ 3.7.2 Conditional Contracts

The conditional contract is one that contains provisions that excuse performance specifically. The contractor does not need to rely on the law or the exceptions to excuse performance. The contractor relies on the contract document itself. However, in order to excuse his or her performance, the contractor must show that the excuse put forward falls within the scope of the excusing provision.

In *Gutman v. Sal-Vio Masons, Inc.*, 72 Misc.2d 729, 339 N.Y.S.2d 562 (Bronx Co. 1972), aff'd 45 A.D.2d 988, 359 N.Y.S.2d 766 (1st Dept. 1974), Gotham Builders Supply sought to recover the cost of construction materials delivered to Sal-Vio Masons. The sale agreement contained a provision that the failure by Gotham to make deliveries because of fire, flood, riot, strike, lockout, labor scarcity, car shortage, freight embargo, government regulation, act of God or the public enemy, legal proceeding, nonarrival or delay of a carrier, or for any other cause related or unrelated over which Gotham had no control would excuse performance. The clause continued that performance would be excused because of the failure, refusal, or inability of any supplier or manufacturer with whom Gotham had contracted for the materials covered by the sale agreement to manufacture, obtain, or deliver the same for any cause.

Gotham had agreed to sell, and Sal-Vio to purchase, all masonry materials required by Sal-Vio. Gotham had also agreed to store certain quantities of gypsum block for the exclusive use of Sal-Vio. Gotham obtained its gypsum block from United States Gypsum Company. When United States Gypsum refused to continue to extend credit, Gotham was forced to discontinue operations.

Gotham, therefore, contended that its failure to provide materials in accordance with the sale agreement did not constitute a breach of contract because its inability to continue in business and obtain further gypsum block relieved it of any further obligation to supply such material to Sal-Vio. The court found that the provision, which excused Gotham for any cause "over which [Gotham] has no control...," was broad enough to protect Gotham from liability in the event it was forced, because of its inability to obtain supplies, to discontinue operations. The court noted that it may not have been the purpose of the clause to excuse nonperformance founded on financial difficulties. The language, however, precludes a finding of an express or implied obligation on the part of Gotham to continue in business.

Green Island Contracting Corp. v. State, 117 Misc.2d 435, 458 N.Y.S.2d 828 (Ct. Cl. 1983), appeal denied 99 A.D.2d 330, 473 N.Y.S.2d 55 (3rd Dept. 1984), aff'd 66 N.Y.2d 605, 498 N.Y.S.2d 1025 (1985), involved the construction of a portion of Interstate Route 88. Work on the project was suspended when a federal court preliminary injunction was issued. The federal action was commenced due to an alleged failure by the State to file an environmental impact statement required by the National Environmental Policy Act. The injunction prohibited the State's receipt and disbursements of federal funds for the project, but did not restrain further construction.

The court found sufficient proof to support a finding that the State's own acts or omissions contributed to the issuance of the preliminary injunction. As early as nine months prior to the letting of the contract, the State knew, or should have known, that the project was not processed in compliance with NEPA. The State ultimately suspended work on the project and unilaterally terminated the contracts. The State argued that its default is excused on the grounds of impossibility of performance. The State argued that since it was restrained by a judicial order it should be excused from performance. However, where the party seeking to be excused has contributed to the issuance of the order, impossibility of performance will not be recognized. Since the injunction was temporary, the court could not find the suspension or interference was so material and substantial that the State ought to be discharged of its obligations under the contracts.

§ 3.7.3 Implied Impossibility

The contract itself may imply a circumstance by which a person will be relieved from his obligation. In *Buffalo v. Bellevue*, discussed at § 3.7.1, the harsh weather and snow conditions excused performance by the trolley company. Performance was rendered impossible by fault other than that of the trolley company.

A similar condition could be implied by the court in connection with a renovation contract. A renovation contract is dependent upon the continued existence of the building to be renovated. The existence of the building is essential to the execution of the contract. If, because of fire or hurricane, or some cause that is neither the fault of the owner nor the contractor, the building ceased to exist, performance is excused.

§ 3.7.4 Excused Performance

How does excused performance affect the financial position of the owner and the contractor? Impossibility of performance is only applicable to the portion of the contract that has not yet been acted upon. Therefore, an owner cannot be unjustly enriched where a construction contract is only partly performed, and it then becomes impossible to perform future portions of the contract.

Should a contractor have received payment in advance for certain work, subsequent impossibility cannot be raised as a defense when the owner seeks to recover the amount he or she paid to the contractor for work now impossible to perform.

3.8 EXCUSE OF PERFORMANCE

§ 3.8.1 Destruction of Building

Once you have a contract to construct a building for an agreed-upon price, you bear the loss of the destruction of that building prior to completion. That is why you maintain comprehensive builder's insurance. How else can you avoid this loss? By explicitly providing in the contract that under particular circumstances the loss will not be yours. Note that the destruction of the building does not create an impossibility that will justify nonperformance.

The issue is raised, with an interesting twist, in *Foley v. Manufacturers & Builders' Fire Ins. Co.*, 152 N.Y. 131 (1897). Foley had constructed foundations for houses being built on his land. Foley engaged a builder to complete the houses. A fire occurred before the time the builder's contract was to be completed. The insurance company argued that the builder was bound by its contract, notwithstanding the destruction of the buildings by fires. Therefore, Foley would not be bound to pay for the work done or materials supplied up until the time of fire. In essence, Foley was not damaged. The insurance company, therefore, resisted payment.

The court noted that the insurance policy protected Foley for loss by fire to the described property. The loss or damage was to be ascertained according to the actual cash value of the property at the time of the fire. The court found that Foley had an insurable interest at the time of the fire. Foley owned the land upon which the structures were being erected. Foley himself had constructed the foundations of the buildings. The insurance company was compelled to admit that the loss sued upon was exactly within the terms of the insurance policy.

Although it was the contractor who would suffer the loss here, the court noted that the relationship between Foley and his contractor was a matter with which the insurance company had no concern.

In fact, these the improvements on the land may cost the owner nothing. When the improvements are destroyed by fire, the owner may compel the builder to replace them without expense. The owner may also recoup by resort to an insurance contract with a third person. In no way do these facts affect the liability of the insurance company absent an explicit exemption in the policy. The court found for Foley.

H.G. Vogel Co. v. Reinhardt, 91 Misc. 60, 154 N.Y.S. 260 (App. T. 1915), aff'd 171 A.D. 907, 155 N.Y.S. 1113 (1st Dept. 1915), involved equipping Reinhardt's factory with a fire extinguishing system. The agreement expressly provided that if Vogel's work should be discontinued by fire or other cause not Vogel's fault, then Reinhardt was to pay an amount equal to the value of the materials, labor, and services furnished to that date. Title would remain with Vogel for the materials and equipment until Foley paid for them.

Before any of the equipment was actually attached to the building, Reinhardt's factory was destroyed by fire. The court focused on the word "services" in the agreement. It held that Vogel should not be limited to the value of the labor performed in the actual annexation of the apparatus to Reinhardt's building. The court found it fair that the value of services should include the cost of surveying the property in preparation to the drawing of plans, securing the details, and preparing drawings of a sprinkler system. The court concluded from the language used by the parties that it was their intent that should the contract be discontinued because of a fire, then Vogel would be reimbursed for the value of his services.

§ 3.8.2 Destruction of Property

Think here of renovation projects. As a contractor, how are your obligations different? Those obligations depend on whether you are providing a complete building or merely performing a task on a portion of an existing building.

Hayes v. Gross, 9 A.D. 12, 40 N.Y.S. 1098 (3rd Dept. 1896), aff'd 162 N.Y. 610 (1900), involved Hayes, a carpenter who agreed to provide interior carpenter and joiner work on a large brick hotel. Hayes was to be paid upon the architect's certificates. During the performance of Hayes' work, the building was destroyed by fire. The fire was without the fault of either Hayes or Gross, the owner. Hayes sought the value of his labor performed and materials furnished. The court noted that when a builder agrees to erect and complete an entire house, and that house is destroyed by fire before completion, the loss falls upon the contractor. If the contractor can still erect another house and does not do so, he or she is guilty of breach of contract.

The court distinguished that situation from one in which a painter agrees to paint a certain house. If the house is destroyed by fire, it is impossible for the painter to complete the contract. Even if the new house is erected, it is not the house the painter agreed to paint. The court asks: Why should the owner pay? Every stroke of the painter's brush converts the painter's labor and material into the property of the owner. A fire thus destroys the owner's property, not the painter's. However, if a painter is painting a boat, which he agreed to make and deliver, and fire destroys the boat before delivery, the whole loss would be the painter's. The boat does not become the property of the purchaser until delivery.

The court noted that in this case, Gross owned the building and Hayes was under a contract to do specified carpenter and joiner work. Hayes was excused from future performance because it became impossible, without his fault, to continue the work. He was entitled to recover for all of the labor and material he added to the building. The fire destroyed Gross' and not the contractor's property.

The court noted that there are numerous cases holding that when the performance of a contract depends on the continued existence of its subject matter, or other essential element, the law implies that the contract is based upon such continued existence. When existence ceases, the obligation ceases. Neither party has recourse against the other on account of breach of contract. The court distinguished these cases and noted that they simply denied a right to recover for damages for breach of contract. The damages were the damages flowing from nonperformance of the contract. In such cases, there will be no recovery because the law implies a condition of continued existence.

In *Hayes v. Gross*, however, an innocent contractor has added labor and material to an owner's property. The nonexistence of the subject matter excuses the contractor from future performance. However, the owner must bear the loss and pay for the labor and material, which he acquired. The owner actually benefited from the services of the contractor. The court notes that other states, while denying the right to recover for nonperformance damages when performance is defeated by destruction of the building, affirm the right of a contractor to recover for what he has done and furnished up to the time of destruction. The court cited *Butterfield v. Byron*, 153 Massachusetts 517; *Cook v. McCabe*, 53 Wisconsin 250; *Haynes v. Second Baptist Church*, 57 Am. Rep. 413 (Maryland).

The contractor was entitled to the value of the labor and materials furnished and his proportionate profits to the time of the fire. The contractor was not entitled to the cost of materials he had procured in order to install them in the hotel since they were not actually in place at the time of the fire. They had not been added to the building and had not become the owner's property.

Remember, the risk of loss can be compensated for by insurance. In addition, specific exclusions may be contained in your contract provisions. The courts are aware of these mitigating factors and will note them when allocating loss on a contract. In *Rhodes v. Hinds*, 79 A.D. 379, 79 N.Y.S. 437 (4th Dept. 1903), Rhodes agreed to manufacture into lumber the logs that a lumber company cut in the vicinity of his mill. Rhodes was to saw and pile the lumber in his yard and load it onto cars when he received direction from the lumber company. Rhodes completed manufacturing and piling the lumber. Approximately half of the lumber had been shipped when a fire destroyed the remainder.

The lumber belonged to the lumber company and was stored in Rhodes' yard for its convenience. Rhodes seeks to recoup the value of the labor actually expended in manufacturing and piling the lumber destroyed. The lumber company owned the logs. The lumber manufactured from those logs was also owned by the lumber company. The company furnished the logs to Rhodes. Rhodes enhances the value of the logs by his labor. Without Rhodes' fault, he is prevented from completing his contract. With the destruction of the lumber, the court holds that the lumber company has failed to supply Rhodes with lumber to ship. The lumber company could have protected itself by insurance. Rhodes, on the other hand, did not own the logs nor the lumber and had no insurable interest. Rhodes could not perform because the company did not possess the property; that is, the lumber essential to render Rhodes' performance possible.

The lumber company argued that Rhodes should have protected himself against the loss or destruction of the lumber by fire by appropriate clauses in the contract. The court rejected this argument. To be sure Rhodes undertook unqualifiedly to deliver the lumber and to complete his agreement before it came due. However, before Rhodes could perform, he needed to have lumber to deliver. The court notes that the lumber had remained piled and that the shipping of the lumber was at the direction of the lumber company. The court found that the necessity for this direction to Rhodes limited the absolute undertaking to ship the lumber. Such would have been within the contemplation of the parties when they made their agreement.

The interpretation is a two-edged sword. It allows Rhodes to collect for labor and services in manufacturing the lumber from logs and piling the lumber. It also relieves Rhodes from liability from failure to perform in shipping the lumber. At the same time, however, it prevents him from recovering beyond the value of the labor and services applied to the lumber company's property. His potential profits on the shipping of the destroyed lumber is lost.

§ 3.8.3 Termination of Existence

Unless your contract provides otherwise, the existence of the object contracted for is necessary for performance. Once that object ceases to exist, without your fault, the court will imply a contractual condition that excuses performance. *Nitro Powder Co. v. Agency of Canadian Car & Foundry Co.*, 233 N.Y. 294 (1922), involved the sale of TNT. The contract contained an express provision excusing performance. After Canadian Car had delivered one carload of TNT, the government of the United States took all remaining TNT by government order. The taking was legal, pursuant to the

National Defense Act. The question is whether or not the contract came to an end. The contract specified that Canadian Car was only to deliver such TNT as could be reclaimed from Russian shells. It further provided that should the quantity be reduced by fire, explosion, government taking over of the materials, or any other cause beyond the control of Canadian Car, then Canadian Car would not be called upon to replace the shortage. The court noted that a contract that is dependent upon the possibility of its performance on continued availability of an item comes to an end when availability ends. When the ending of the availability is beyond the control of the parties, the contract is regarded as dissolved. By its terms, the contract was conditioned on government inaction. The taking by the government frustrated the contract because it left no material upon which the contract could operate. Once the material was taken, Canadian Car was released from all obligation to Nitro Power.

§ 3.8.4 Contractor Contributes Only Part

A contractor providing labor and services in a renovation project or in a portion of a larger work is excused from performance when the building or project on which he is working is destroyed prior to completion. The case of *Hayes v. Gross*, discussed in detail at § 3.8.2, is the best illustration of this principal. Where the owner has control of the building, the liability for destruction of the building is his.

§ 3.8.5 Cessation of Subject Matter

Sheer nonperformance of a contract will be excused where without your fault, you are incapable of performing the contract because an object essential to the contract ceased to exist. Some thought must be given to express contractual provisions allowing for various contingencies. Many courts will hold that a circumstance that could have been known or anticipated at the time of the making of the contract should be accounted for in that contract. If you could have guarded against a certain contingency and you did not, the courts can hold you to the contract. Of course, as we have seen, the courts can often imply conditions in a contract. The safest route, however, is to consider those contingencies that may affect your performance and to provide for them explicitly in the contract documents themselves.

§ 3.8.6 Intention of Parties

Is your performance excused? While the courts may imply conditions for excusing performance, the surer course is for you and your attorney to review contingencies and specifically incorporate them in the contract. The courts then simply interpret the contract to see if the offending occurrence was within the contemplation of the parties and stated as excusing performance.

The performance of a construction contract necessitates your dealing with materialmen. Those dealings, too, are subject to various contractual terms. What are the consequences of government intervention in connection with a contract for delivery of materials? This topic was touched upon in the *Nitro Powder* case, discussed at § 3.8.3.

J.H. Labaree Co. v. Crossman, 100 A.D. 499, 92 N.Y.S. 565 (1st Dept. 1905), aff'd 184 N.Y. 586 (1906), involved the sale of coffee. Once the coffee had arrived in New York Harbor, the Board of Health prevented its unloading. The court notes the general rule that an undertaking by a party is *not* excused because some unforeseen contingency has rendered the performance impossible. The rule protects the integrity of contracts since the party could have provided for the contingency in the contract itself. There are, however, numerous exceptions. If the law interferes and renders performance impossible, a party is excused. Persons must always be considered as contracting with reference to the law existing at the time of contract. If the existing provisions of law render performance impossible, the contract is deemed dissolved. The court noted that in these cases, it is not essential that the contract contain provisions granting relief from performance.

The court noted that in this case the contingency that the coffee would be absolutely excluded from the city was a remote one. It was undoubtedly not in the minds of the contracting parties. Crossman had done all he could legally do. The Board of Health's intervention legally excused performance of the contract.

§ 3.8.7 Inevitable Accident

The inevitable accident or act of God is in these more secular times denominated by the Old French, "force majeure," or, roughly translated, major impact. A force majeure provision will typically excuse performance because of fires, floods, earthquakes, other acts of God, acts of labor unions, strikes, war, and "other conditions beyond the control of the parties."

Why then is there case law constantly seeking excuses of performance? The contracting parties simply neglect to consult their attorneys and consider foreseeable contingencies that would impede or prevent performance. Those contingencies should be considered and spelled out in the contract documents themselves.

Ogdensburg Urban Renewal Agency v. Moroney, 42 A.D.2d 639, 345 N.Y.S.2d 169 (3rd Dept. 1973), involved an urban renewal project. Ogdensburg began condemnation proceedings against Moroney. The parties agreed that the property would be purchased for $100,000 and reduced this agreement to a writing in the form of an order of condemnation. Ogdensburg later discovered that the Federal Department of Housing and Urban Development would set a maximum value of $93,000 for the property and would not approve the $100,000 purchase nor supply the funds its acquisition. The court found that this development did not excuse Ogdensburg from its obligation to pay Moroney $100,000. Without an express qualification or provision in the contract, this occurrence cannot discharge the contractual obligation. Ogdensburg argued that performance had become impossible due to a change in the laws. The court rejected this exception as inapplicable. Ogdensburg had the authority to make the contractual commitment it had made without prior federal approval. The fact that federal funds on which Ogdensburg had relied to purchase the property are not available does not render performance impossible. Ogdensburg has other statutory sources for available funding.

While the unavailability of federal funding presents Ogdensburg with additional difficulties and inconveniences, it does not render performance impossible. The defense of impossibility is only available where performance is rendered impossible by the happening of an unanticipated event, which could not be foreseen or guarded against in the contract. Ogdensburg knew at the time of the agreement of sale that it would need federal approval if federal monies were to be made available for purchase. However, the agreement is silent concerning the obtaining of this approval. The court found for Moroney.

Moyer v. City of Little Falls, 134 Misc.2d 299, 510 N.Y.S.2d 813 (Herkimer Co. 1986), involved a public contract for refuse disposal. A factor that influenced the amount of Moyer's bid was his ability to dump refuse in a landfill near the City of Little Falls. During the contract term, that landfill was closed by order of the New York State Department of Environmental Conservation. The action by the Department of Environmental Conservation created a monopoly. The cost to Moyer per cubic yard of refuse increased by almost 700 percent. Moyer argued that he should be discharged from his contract due to impossibility of performance. The court analyzed Moyer's position and found that he was doubly blessed. Performance of the contract was financially impossible as a result of actions of a government agency. While increased costs alone will not excuse performance, a rise in costs due to an unforeseen contingency, which alters the essential nature of the performance, may excuse performance. Where conditions are so drastically altered, the courts can find that reasonable persons would not have made the subject contract. Where an unanticipated circumstance has made performance vitally different from what should reasonably have been within the contemplation of both parties, the risk should not unfairly burden one of those contracting parties.

The court recognized a growing trend that performance should be excused (1) if government or other contingencies create a substantially unjust situation totally outside of the contemplation of the parties, and (2) that the situation is such that an experience draftsman could not have reasonably anticipated it at the time of contract signing. While the court finds that the sanitary landfill problem was a major public concern, and enforcement efforts by the Department of Environmental Conservation were foreseeable, the creation of a monopoly was not. It was unreasonable and unfair to require Moyer to bear the brunt of the Department of Environmental Conservation's action. The parties could not anticipate such a result. Moyer is excused from further performance.

A force majeure provision can be interpreted broadly. Excuse will not be found by a court, however, unless the occasion of that excuse is specified in the provision.

Is weather an act of God, which would excuse performance? In *Buffalo v. Bellevue*, discussed at § 3.7.1, weather was considered an implied condition to performance. In these more sophisticated times, most force majeure provisions will include weather as a specific excuse. The impact of weather has historical implications. Many contracts require that the records of the National Oceanic and Atmospheric Administration be consulted for a number of prior years to determine just what are the normal weather conditions at the project site. A blizzard in southern Florida is not the same as one along the route of the Alaska pipeline. In any event, a claim for days of delay due to weather conditions should always deduct the number of days that a reasonable contractor would have anticipated experiencing during the term of construction.

Is rain in a wet season of a wet state "severe weather?" *Roger Johnson Constr. Co., Inc. v. Bossier City*, 330 So.2d 338 (C.A. Louisiana 1976), involved the construction of recreational facilities at three park areas. The contract was completed 45 days after the contractual completion date and Bossier City assessed liquidated damages. Johnson seeks return of the amount held for liquidated damages. The contract did provide that Johnson would not be charged with liquidated damages when a delay in completion is due to unforeseen cause beyond his control and without his fault or negligence including acts of God, public enemy, acts of the owner, acts of another contractor, fires, floods, epidemics, quarantine restrictions, strikes, freight embargoes, and severe weather. In its testimony, the contractor pointed out that the impact of weather conditions went far beyond mere interference by falling rain. The water content of the soil would not permit the laying of asphalt, nor of preliminary soil preparations. Johnson could not work because heavy rains delayed soil tests. Relying on historical data, an officer of the National Weather Service testified on Johnson's behalf. Over the contract period normal rainfall would have been 30.99 inches. There was a plus departure of 2.82 inches. Bossier City's inspector testified that there were only 38 days during the entire period of the contract when Johnson was unable to proceed with the work. The City's civil engineer also testified, and his inspection and the inspection reports also showed 38 days lost due to wet weather. The court found that the cause of the delay was lack of sufficient personnel. The project could have been timely completed by Johnson using sufficient crews on all sites on the days during which work could have proceeded. The court upheld Bossier City's withholding of liquidated damages from Johnson.

Where weather data supplied by the contractor does not demonstrate higher levels of precipitation that could have been reasonably anticipated, then those weather delays are not excusable. *McDevitt & Street Co. v. Marriott Corp.*, 713 F.Supp. 906 (E.D. Virginia 1989). Remember that the courts view severe weather as that which could not be anticipated by the parties to the contract. Therefore, excusable delay results from those weather conditions that substantially deviate from the historical record.

§ 3.8.8 Strikes

As a general rule, the courts do not consider strikes an unanticipated event excusing performance. Should you wish the occurrence of a strike to be considered by the court, it should be explicitly placed in your force majeure provision. As always, the strike must be a proximate cause of the breach in order to excuse performance under the provision.

Barnum v. Williams, 115 A.D. 694, 102 N.Y.S. 874 (2nd Dept. 1906), aff'd 190 N.Y. 539 (1907), involved a contract in which Hecla Ironworks agreed to fabricate and erect for Williams certain light and ornamental ironwork on his premises. Hecla proceeded as far as it was able, but the owner made certain changes to the plans. It was not until ten days after the time when Hecla, by the terms of the contract, was to have been completed, that Williams sent the revised plans. It was undisputed that Hecla would have been able to complete its contract by the contract completion date had the plans been received sufficiently in advance. Approximately two weeks after the plans were received, Hecla's employees struck the work. The court found that Hecla Ironworks was excused from completion of the contract by the strike. Where a contractor is delayed in the completion of the contract by fault of the owner, as here, his or her time to complete is thus extended for a reasonable time.

In general, the extension of a contractor's time does not include any period, long or short, during which he or she is unable to proceed by reason of a strike. In *Barnum*, the court found that the contractor had been delayed by the owner for a period far beyond that necessary to complete and far beyond the time within which he could have easily completed. The result must be different. If the contractor proceeds and does not rescind the contract, then the reasonable time within which he or she is allowed to complete does not include the period of time within which it is utterly impossible for him or her to proceed. The employees of Hecla who struck the building were all members of the same union. Hecla was quite as effectually prevented from proceeding as though prevented by acts of mobs and violence.

Munson v. Smith Woolen Machinery Co., 118 A.D. 398, 103 N.Y.S. 502 (3rd Dept. 1907), involved a contract by Smith to provide certain manufacturing equipment for a knitting mill. The contract specifically provided that Smith would not be liable for delays caused by strike. The court found that the contract contemplated final furnishing of sufficient adaptable machinery. Smith failed to perform its part of the contract and, therefore, cannot rely on the provision that exempted Smith from liability by reason of strikes. When a contractor utterly fails to perform a contract, he or she cannot rely on such a provision to hold him or her safe from liability.

Finally, *McGovern v. New York*, 234 N.Y. 377 (1923), reh'g denied, remittitur am'd 236 N.Y. 508 (1923), illustrates the courts' approach to strikes. McGovern had contracted with the City of New York to construct a subway for a lump sum price. His contract allowed, as public contracts often do, for extensions of time due to strikes.

World War I intervened. Men were siphoned off to the front and to various war related industries. The approximately 6,000 workmen on the subway project threatened to strike unless their pay was increased. McGovern stated he would agree to a pay increase if the City would undertake to pay the extra cost. McGovern instituted a new pay scale. When the City refused to enter into any binding agreement, the workmen struck the project. The City then agreed to pay any extra costs. The City subsequently refused to pay on the agreement. McGovern sought to enforce the agreement with the City.

The court found that the contractor cannot invoke the strike provision. The strike was due to the failure of McGovern to pay wages essential to keep the work going. If cheaper labor cannot be obtained at the prevailing market rates, if the wages are reasonable in the amount, and if the contractor refuses to pay those wages, then the strike is caused or provoked by the contractor. The strike provision does not entitle a contractor to abandon all effort to avert the strike whenever a strike is threatened. The City was entitled to receipt of its subway at the lump sum price. To pay for the extra work is construed by the court to be a gratuity, forbidden by the state constitution. The risk of an increase in wages over the course of the contract is a risk assumed by the contractor.

§ 3.8.9 Legal Changes

Where performance of a contract becomes impossible because of a change of the law or an action taken under government authority, performance is excused. The principal is simple. The contract was made consistent with the law at the time of its making. Because of a change in the law, it has subsequently become illegal. An implied condition of a contract is that a party to the contract will not be compelled to perform an illegal act.

Philadelphia Boiler Works v. Foundation Co., 190 N.Y.S. 696 (App. T. 1921), involved a change in position by the government. Foundation Co. had awarded to the boiler works a subcontract to furnish and erect one steel smoked breeching. Foundation Co. subsequently notified the boiler works not to proceed with the order. Foundation argued that it had made a contract with the United States, then at war, which required it to construct and ammunition plant for the government. The breeching was to be a part of that plant. The contract was of a standard form used by the government in national emergencies. It recited that the work was urgently required and performance must be completed within the shortest possible time. The contract was subject in every detail to orders of the Contracting Officer of the United States. Without the Contracting Officer's authority, no subcontract could be made. The Contracting Officer was empowered by the contract to make changes to the work, issue additional instructions, and to direct omission of work previously ordered. The subcontract given to the boiler works had the approval of the Contracting Officer. The Contracting Officer decided that the breeching would be absolutely necessary within three weeks. The boiler works did not have the material in stock. Cancellation of the subcontract by Foundation Co. was made at the express direction of the Contracting Officer. At the time Foundation Co. and the boiler works made the subcontract, the boiler works had knowledge of the terms of the Foundation Co. contract with the United States. The court found that both parties contemplated the contingency that performance might not be permitted by the United States. It was an implied part of the subcontract. If performance was prevented by the United States, both parties were released from subsequent obligation. Government intervention will excuse both parties from further performance without giving either the right to recover damages for the part not performed.

What must the involvement of the government be? *Standard Oil Co. v. Central Dredging Co.*, 225 A.D. 407, 233 N.Y.S. 279 (3rd Dept. 1929), aff'd 252 N.Y. 545 (1929), discussed at § 3.4.2, involved a dredging contract. The dredging company contracted to deposit on Standard Oil's land a large quantity of spoil from a government contract on which it was working. At the time of contracting, Central Dredging knew that permission of the United States division engineer was necessary to deposit the spoil upland. Central Dredging asserted that its contractual obligation was extinguished when the division engineer refused permission. The court found that the refusal by the division engineer did not of itself abrogate the contract. Central Dredging failed to show sufficient diligence in its effort to obtain permission. The determination of the division engineer could have been reviewed by the chief engineer of the Corps of Engineers and the chief engineer's determination reviewed by the Secretary of War. Difficulty, or even improbability of accomplishment, will not release the contractor. The government's failure to grant the permit does not establish a defense to the contractor. For a defense to be established it must be shown that the object of the contract cannot by any means be effected.

So, too, in *Ogdensburg Urban Renewal Agency v. Moroney*, 42 A.D.2d 639, 345 N.Y.S.2d 169 (3rd Dept. 1973), discussed at § 3.8.7, the urban renewal agency was not released from its contract to purchase property. The agency had agreed on a price. The federal government determined that the maximum value of the property was below the selling price and would not approve purchase at the selling price, nor supply funds for its acquisition. The fact that federal funds upon which the agency relied to purchase the property are not available does not render the performance of the purchase contract impossible. The government action does not provide a defense to the contract. The court found for the seller.

§ 3.8.10 Insolvency

Most construction contracts will contain a provision releasing the contractor should the owner become insolvent, bankrupt, or have a receiver appointed. The general rule is that mere insolvency does not release another party. Again, this is a contingency that should be dealt with in the contract document. However, if one party's performance is based on subsequent performance by the other party, and the other party is unable to perform because of insolvency, the first party's obligation is at an end. See *Updike v. Oakland Motor Car Co.*, 229 A.D. 632, 242 N.Y.S. 329 (1st Dept. 1930).

In *Pettinelli Electric Co., Inc. v. Bd. of Ed.*, 56 A.D.2d 520, 391 N.Y.S.2d 118 (1st Dept. 1977), aff'd 43 N.Y.2d 760, 401 N.Y.S.2d 1011 (1977), Pettinelli agreed to construct two schools. The chief need for the schools arose from an anticipated housing project, which the State of New York Urban Development Corporation was planning to build. Because of the Urban Development Corporation's fiscal difficulties it decided not to build the housing project. The Board of Education canceled Pettinelli's contract and terminated construction for the schools. Pettinelli sought damages for breach of contract.

The Board argued that the state legislature had declared that a financial emergency existed in the City of New York. This emergency also affected the Board of Education. To cope with the financial emergency, the Board of Education had terminated and canceled the contracts. Therefore Pettinelli, the Board argued, may not maintain the action.

The court noted that the defense is insufficient in law. Financial difficulty, economic hardship, even insolvency or bankruptcy do not establish impossibility so to excuse the Board from liability in damages. The housing project was not so intimately related to the schools that it may be fairly said that the discontinuance of the housing project legally frustrates the school construction contract. The decision by the Board not to continue with the school construction contract is analogous to a business decision that would render the argument of frustration inapplicable. The court found for Pettinelli.

In *Stannard v. Robert H. Reid & Co.*, 114 A.D. 135, 99 N.Y.S. 567 (1st Dept. 1906), later appeal 118 A.D. 304, 103 N.Y.S. 521 (1st Dept. 1907), aff'd 195 N.Y. 530 (1909), the court carefully analyzes the effect of a voluntary dissolution. Stannard had a contract with the United States government to erect a courthouse and post office. Stannard had subcontracted with Reid to do marble work and lay the floors at the post office. Before the construction arrived at the stage requiring that Reid's work begin, the corporation was voluntarily dissolved and temporary receivers appointed by the court. The receivers were authorized to complete contracts of the corporation. However, they gave notice to Stannard that because his subcontract was an undesirable one, likely to result in a loss, the contract would not be completed. Stannard advised the receivers that he considered this notice to be a default and that he would relet the contract and hold Reid for any excess of cost. One year after the receivers were appointed, the court discharged them and vacated the order appointing them. This court found that Reid was not, as a matter of law, prevented from performing the contract in question by the appointment of receivers. Nor did the appointment relieve Reid from liabilities for damages in breach.

With respect to the voluntary dissolution of a corporation, whether or not that dissolution relieves the corporation of its liabilities on a contract is a question of fact to be determined upon consideration of all the circumstances. If the corporation is voluntarily dissolved for the purpose of avoiding undesirable contracts and not because of necessity, that act is a fraud. The corporation is estopped from setting up such a dissolution as a defense to a breach contract action. If dissolution is a necessity, a corporation is generally relieved of its contracts.

The courts point out that where a person seeking damages in breach of contract has himself or herself instigated the dissolution and was a reasonable cause of it, that person could not assert that it was not a necessity. The court in *Stannard* noted that within a few days of their appointment, the receivers obtained authority to carry out contracts of the corporation. They presumably proceeded to carry out those they had deemed profitable and repudiated those that the deemed unprofitable. After the receivers had supposedly relieved Reid of its obligation under the unprofitable contract with Stannard, the property of the corporation was restored to Reid. The receivers were then discharged and the order appointing them vacated. This state of facts raised a fair presumption that the financial condition of the corporation did not make its dissolution a necessity. Rather, the dissolution was a temporary expedient for the purpose of relieving Reid from an undesirable obligation. It was not enough that Reid showed it had been voluntarily dissolved. It was necessary for Reid to go further and show that the dissolution was a necessity because of its financial condition. This Reid did not do. The court found for Stannard.

§ 3.8.11 Death

As a general rule, where a contract requires personal service, the death of the person who is to supply that service excuses performance. Performance is impossible by reason of the death of such person. The courts imply a condition in the contract that death dissolves the contract. From the nature of the contract, it should be evident that the parties to the contract rely on the continued existence of a particular person. *Buccini v. Paterno Constr. Co.*, 253 N.Y. 256 (1930), involved a contract with Paterno to decorate a ballroom, banquet hall, and swimming pool of a dwelling. The character of the decorations was to call for the exercise of artistic skill. The contract explicitly provided that all decorative figured work was to be done by Buccini personally. Only plain work was to be delegated to his assistants.

Buccini died while the work was in progress. The court found that since the contract was personal, the affect of death was to terminate the duty of going forward with performance. However, the owner was liable for the benefits received.

§ 3.8.12 Blockade

The question of blockade rarely arises in construction contracts. It can, however, become a concern where machinery or instrumentation is being imported. Again, if you are importing foreign machinery, a contract provision should be included to excuse performance because of a blockade or an embargo. The case of *Thaddeus Davids Co. v. Hoffman-La Roche Chemical Works*, 178 A.D. 855, 166 N.Y.S. 179 (1st Dept. 1917), is illustrative of the applicable principles. Hoffman prepared a contract that was signed by Davids without change for the sale of carbolic acid crystals. The contract stated that contingencies beyond Hoffman's control (fire, strike, accident, or change in tariff) would allow Hoffman to cancel the contract. After two deliveries, Hoffman flatly refused to furnish any more crystals. It stated that the European governments whence it obtained its supply had placed an embargo on exportation. None had been received from Europe since the outbreak of the war and Hoffman's stock had been exhausted. Hoffman considered this "contingencies beyond our control" and canceled our contract. The court found that Hoffman, who had offered the contract, must be bound by it. Hoffman had failed to provide against the contingency of foreign wars and embargoes by foreign powers. The court found that the cancellation by Hoffman was unjustified. Hoffman was liable for damages.

§ 3.8.13 Third Person's Actions

Suppose the cooperation of a third party is necessary for performance. The contractor's inability to control that third party cannot give rise to an impossibility that excuses performance. In *De Long v. Zeto*, 135 A.D. 79, 119 N.Y.S. 765 (1st Dept. 1909), De Long was to furnish and deliver to Zeto certain building materials for the construction of four frame houses for Cosmos Realty Company. While De Long was to furnish and deliver the material, it was not to put any of it in place. He was to receive his first payment when standing trim was up, his second when the buildings were complete, and his third, 30 days thereafter. The evidence showed that De Long had delivered all the materials and that they were accepted by Zeto. The first installment was paid, but the other two were not.

After delivery of the last materials, the mortgage on the premises was foreclosed. The action for the last two payments was commenced nearly one year after De Long completed delivery of the materials. If the houses were then complete, there is no defense to De Long's lawsuit. If they were not complete, then Zeto had failed to complete them within a reasonable time and De Long is still entitled to recover. The failure to complete buildings owing to foreclosure of a mortgage is no defense. De Long was not responsible for the foreclosure action. Its consequences cannot be visited upon him. The court finds for De Long.

§ 3.8.14 War

Again, the best advice is to have a contract provision that deals specifically with war. In the absence of a provision, however, the courts will excuse performance where that performance is wholly prevented by war. This is, of course, the impossibility argument. As the next case indicates, the courts are not always willing to acknowledge impossibility.

North Hempstead v. Public Service Corp., 107 Misc. 19, 176 N.Y.S. 621 (Nassau Co. 1919), aff'd upon opinion below 192 A.D. 924, 182 N.Y.S. 954 (2nd Dept. 1920), involved a franchise which the town of North Hempstead granted to Public Service to maintain and operate gas mains and pipes through the public highways of the town. Such a franchise, when accepted, becomes a contract between the parties. North Hempstead brought an action against Public Service to recover damages for nonperformance. As a defense to nonperformance, Public Service interposed that the activities of the War Industries Board controlling iron and steel production rendered performance impossible. The court noted that the rules and regulations of the War Industries Board greatly increased the difficulty, and probably the expense, of securing pipe. However, such difficulty is no different than if some other unforeseen contingency such as scarcity of labor, strikes, or increased taxation restricted the supply. The controlling consideration is that neither the acquisition nor the use of the pipe for the purpose of fulfilling the obligation assumed by Public Service was forbidden or rendered illegal. Official actions caused by the war may have increased the difficulty of performance, but this is not enough. Increased difficulty and expense of performance never excuses performance. Whether an application to the War Industries Board for a priority order would have been granted or refused, or whether the rules and regulations did or did not provide for a priority order, are considerations that beg the question. The power to grant a priority order was at all times in the Priorities Board or the War Industries Board. If in the exercise of wise precaution, Public Service had previously acquired a surplus supply of pipes, the use thereof was never prohibited nor rendered illegal. Neither was acquisition of pipe from private sources prohibited. The court noted that all that can be said is that Public Service was confronted by an unforeseen contingency might have been provided for in the contract. Not having done so, Public Service is liable for performance. The court found for North Hempstead.

§ 3.8.15 Government Control

If the means of performance of the contract are placed under government control, performance may be excused. This is exactly the issue discussed in *Nitro Powder* at § 3.8.3.

Empire Lumber Co. v. Parshelsky Bros., 201 A.D. 764, 194 N.Y.S. 670 (2nd Dept. 1922), involved a dispute arising at the conclusion of World War I. Empire purchased twelve cars of lumber. The material was solely for use of the United States in connection with military operations. No freight car could be placed for use for such transportation without the consent of the government, evidenced by permits issued either at Camp Upton or by a bureau in Washington. Parshelsky, however, failed to deliver the permits. Parshelsky conceded that the duty for procuring the permits was his. He furnished no permits for an extended period and then permits for only three of the twelve cars. By this time, Empire had notified Parshelsky that it had canceled the order. The court noted that Parshelsky had undertaken something that it was powerless to procure. The only question was whether Empire was bound to wait even longer before canceling the order. Empire had accepted the order, subject to war measures. One of the innumerable war measures was that no car could be moved without the consent of the government. Here, Parshelsky had taken upon himself the removal of any impediment of government intervention. The court finds that the risk fell with Parshelsky, and any injury resulting from the risk cannot be transfered to Empire. The court found for Empire.

3.9 ACT OF ADVERSE PARTY

§ 3.9.1 Nonperformance of Condition

The general rule is that a party to a contract cannot rely on the failure of another to perform a condition precedent where that first party has frustrated or prevented the occurrence of the condition. A person cannot take advantage of a condition precedent, the performance of which he or she has rendered impossible. In *Kooleraire Service & Installation Corp. v. Bd. of Ed.*, 28 N.Y.2d 101, 320 N.Y.S.2d 46 (1971), a public contract stated that it would not be binding unless the comptroller endorsed his certification that there remained an unexpended and unapplied balance of an appropriation sufficient to pay the expense of the contract. Kooleraire had progressed through the preliminary stages of the work. The Board of Education then rescinded the contract. The comptroller had not endorsed his certification on the contract. The evidence established that funds were available for the contract, and that a later contract was certified by the comptroller for this purpose. The Board argued that since the comptroller had not certified the contract, Kooleraire is precluded from any recovery.

The court found that the contractual provision had one purpose: to show sufficiency of money to do the work. The provision should not be utilized to accomplish another purpose. When the funds are available, the document should be certified. The comptroller's failure to certify was not a legal defense to the action.

A similar situation occurred in *McIntosh Ready Mix Concrete Corp. v. R.D. Battaglini Corp.*, 36 A.D.2d 561, 317 N.Y.S.2d 692 (3rd Dept. 1971). McIntosh was supplying concrete to Battaglini for construction of a flood control project. The concrete supplied was required to undergo certain strength tests. The public employees who were to control the test sampling violated various aspects of the testing requirements. The public agency was not justified in rejecting the questioned concrete. Where the work had been substantially performed in accordance with the provisions of a contract, the failure to take or pass stipulated tests due to the fault of the party for whose benefit the tests were to be accomplished does not bar recovery. The owner cannot rely on a condition precedent such as the fulfillment or passing of the test where the nonperformance of the condition was caused by the owner himself.

§ 3.9.2 Refusal to Perform

What facts amount to a refusal to perform? If one party refuses to perform, what are the obligations of the other party? As we have seen, a breach by one party will excuse performance by the other. In *Sherry v. Federal Terra Cotta Co.*, 172 A.D. 57, 158 N.Y.S. 241 (1st Dept. 1916), Sherry's agent had contracted with Federal to supply terra cotta trim or ornamentation to be used in the construction of a country home. Some time after an initial payment had been made by Sherry, he and one of his builders went to Federal's factory to see what had been done with respect to the terra cotta. After Sherry inspected the working drawings, which included a number of nude figures, he asked Federal to substitute conventional urns and plain terra cotta. A few days after the parties had agreed upon the substituted patterns, Federal informed Sherry that it could not proceed on the agreed price, but must have more money. Federal stated that the contract was only the first or second that it had ever entered into. The contract price was too low, and Federal also claimed that Sherry's former architect had written a letter canceling the contract. Sherry expressed surprise at the allegation that the contract had been canceled, but agreed upon an increased amount for the work upon the condition that the architect's letter canceling the contract be delivered to him. By that means, Sherry hoped to compel the architect to make good the extra amount he was compelled to pay. The architect's letter canceling the former contract was never sent. Some time later, Federal admitted that it had never received any instructions from the architect to cancel the contract. Sherry then demanded that Federal carry out the prior contract and deliver the terra cotta. Having received no answer to his letter, Sherry then wrote to Federal and stated that he would place his order elsewhere and expected a refund of the money paid under the original contract. Federal replied and demanded an additional sum to cancel the contract.

Who has breached? The court does not make this decision. It is a question of fact decided by the jury. The court does point out the various issues to be considered. A jury could have found a breach when Federal refused to go on with the work unless it was paid more money, and falsely stated that the prior contract had been canceled. This would have relieved Sherry from any obligation under the contract. That Sherry did not act at once on this breach and order terra cotta elsewhere is not to be deemed an acquiescence. His negotiation with Federal was induced by the false statement that the original contract had been canceled. Sherry, in his negotiations, had relied on that statement. If the jury found it was Federal who repudiated the only contract between the parties, then Sherry is entitled to damages.

In *J.W. Brennan Const. Co. v. State*, 117 Misc. 816, 191 N.Y.S. 253 (Ct. Cl. 1921), Brennan was to improve a state highway. At a certain point, a concrete culvert was to be constructed. The plans at this point noted, "to be built by the town." A dispute developed between the state and the town with respect to the concrete culvert, the town contending that it was unnecessary. Brennan was ignorant of the town's position. The state executed the contract and Brennan proceeded with his work. The court emphasized the general obligation of the State to provide its contractors with a site for the work and to refrain from impeding the performance of the work.

Brennan was unable to continue its work because the culvert had not been constructed. Brennan had been repeatedly assured by the State that it would soon be in place. The failure of the State to perform excuses Brennan's performance. Brennan is not required to spend considerable monies and effort to construct a detour.

§ 3.9.3 Prevention of Performance

There is an implied obligation in a construction contract that the owner will not obstruct the construction, but will facilitate it.

The owner has an obligation not to interfere with the prosecution of the work of his or her contractor. The owner impliedly agrees that the contractor will not be unreasonably delayed by the failure of other contractors to perform work that is essential to the performance of the contractor's work. *Shalman v. Bd. of Ed.*, 31 A.D.2d 338, 297 N.Y.S.2d 1000 (3rd Dept. 1969).

Jones v. Dodge, 137 A.D. 853, 122 N.Y.S. 815 (2d Dept. 1910), involved the construction of a house. Jones received the first four partial payments. The fifth payment, which was due upon completion of the plastering, was not made and was not earned.

The issue was by whose fault the plastering was left incomplete. Jones contended that the owner so interrupted the work that he was not able to complete it. The owner, on the other hand, stated that Jones could have devoted himself to other areas that needed plastering unaffected by the owner's interruptions and failure to have the electrical wiring completed.

While Jones was entitled to earn his payments as delineated in the contract, the owner was not in a position to force Jones to distribute his labor over the various areas of work. Plastering could not be performed until the wiring was performed. Trim work could not be performed until the plastering was performed, and the fifth payment could not be made until the plastering was performed.

The evidence indicated that the owner was preventing the electrician from completing his work. Therefore, Jones could not complete the plastering and earn his money according to the terms of the contract. Jones would be entitled to earn the payments for his work. The owner cannot obstruct him and prevent him from earning an installment.

How does an owner prevent performance? In *Felix Contracting Corp. v. Oakridge Land & Property Corp.*, 106 A.D.2d 488, 483 N.Y.S.2d 28 (2nd Dept. 1984), appeal denied 66 N.Y.2d 606, 519 N.Y.S.2d 1025 (1985), the court found that Oakridge was in breach of its contract with Felix for excavation and site development. Felix was justified in stopping work when Oakridge materially breached the contract. After work commenced, Oakridge issued approximately 70 revised drawings. These resulted in numerous construction changes. Oakridge refused to acknowledge that Felix was entitled to additional compensation. In addition, Felix experienced delays due to Oakridge's failure to obtain required building permits. Oakridge had misrepresented to Felix that the project would be a "balanced job." Material excavated from a lake on the site was to be suitable for use as roadway subbase material. This was not the case, and additional costs were incurred. Oakridge generally failed to coordinate the work on the project. The court found that Felix was entitled to compensation for work performed pursuant to the contract as well as for work performed outside the scope of the contract.

§ 3.9.4 Excusing Nonperformance

The contractor's nonperformance of a condition precedent can be fatal. However, what are your obligations when the owner breaches? You are not required to perform those conditions precedent when it is clear that your performance will not result in performance by the owner. The courts recognize that your performance of the conditions precedent has become futile by the prior breach of the owner. The courts will not consider performance of the conditions precedent to be a prerequisite to your recovery once the owner has breached. The courts will deem that the condition precedents have been waived by the breaching owner.

§ 3.9.5 Refusal to Pay

There are many fine distinctions that result in many fine interpretations by the courts. One issue is clear, however. An owner's refusal to pay relieves the contractor of his or her obligations under the contract. Because of the time span over which most construction contracts are performed, the issue most often comes into play in connection with the prompt processing of partial payments.

Parisi v. Hubbard, 226 A.D. 280, 235 N.Y.S. 220 (4th Dept. 1929), involved a dispute between a general contractor and his plumbing and heating subcontractor. Parisi progressed the plumbing and heating work to the point where he was entitled to receive his first partial payment on the subcontract. The general contractor refused to pay the amount due. Parisi stopped work and declined to go further until he received his money. The court found that this he had a right to do and would not be charged with having breached his contract.

You, as a contractor, must be cautious in determining whether the owner or your subcontractor has breached. The courts do not take kindly to the idea of a party abandoning contractual obligations. If you are found to have insufficient cause for your inaction or abandonment, the court will find that *you* have breached the contract. *Empire Lighting Fixture Co. v. Browning*, 93 Misc. 489, 157 N.Y.S. 284 (App. T. 1916), involved the furnishing, delivering and hanging of certain lighting fixtures and chandeliers. The written contract contained no provision for payment as the work progressed. Payment could not be demanded as of right until completion of the contract. The contract required that the plaintiff was to furnish certificates from the New York Board of Fire Underwriters for each of the fixtures installed. The court found the failure to furnish certificates not merely a technical omission, but one of substance. The lack of a certificate may seriously affect fire insurance and the Department of Water Supply, Gas and Electricity may cause electric current to be discontinued. When Browning refused to make payments, Empire Lighting did not supply all the fixtures called for. The court found for Browning, holding that he was under no obligation to make payments demanded.

In *Sklar Door Corp. v. Locoteta Homes, Inc.*, 33 Misc.2d 299, 224 N.Y.S.2d 294 (Nassau Co. 1961), Sklar had contracted to provide doors and appurtenant materials in each of 25 houses. Locoteta built its first model house, but did not pay for the doors and appurtenant materials until some time after completion. Locoteta built two more homes. Doors and appurtenant materials were again installed by Sklar. Locoteta was late in payment. Locoteta was about to build additional homes when Sklar declared it would not furnish any more doors and materials until it was assured of payment on delivery or installation.

Sklar claimed it was prevented from completing the performance of the contract by Locoteta. The contract drafted by Sklar does not say that payment was to be made immediately upon the installation and delivery to each and every house. The language is that payment is immediate upon completion of installation and/or delivery of the merchandise sold thereunder. The court found that a reasonable interpretation would permit the conclusion that payment was to be made upon the installation or delivery of all 25 houses as well as on each house. Since there is ambiguity in the language of this contract, which was prepared by Sklar, it is resolved against Sklar. The breach of the contract was by Sklar rather than by Locoteta. The court found for Locoteta.

§ 3.9.6 Subscriptions

As a contractor, you will only become peripherally involved with subscriptions. What exactly are subscriptions? Subscriptions are the means by which funds are raised for various charitable purposes, for construction of churches, war memorials, and educational institutions. *Re Metz' Estate*, 262 A.D. 508, 30 N.Y.S.2d 502 (1st Dept. 1941), involved a subscription for the erection of the "Broadway Temple." This was an ambitious project including a 12-story apartment house at each corner of a large lot and, in the center, an enormous tower that was to contain the auditorium of a church and upper floors with hundreds of apartments. The edifice was to include bowling alleys, a gymnasium, social hall, sunday school, and meeting rooms.

The buildings were to operate as a single unit. The revenue derived from the real estate was to be used for the religious, charitable, and social activities of the Temple.

The fund raising campaign commenced, and a number of subscriptions were elicited. Through these private contributions, the sale of second mortgage bonds, and monies raised by a first mortgage, two apartment buildings, and the foundations of the central structure were constructed. At this juncture it was discovered that the buildings would cost 6 rather than 4 million dollars. Additional pledges were subsequently secured.

Work resumed on the uncompleted building. It consisted of completing an auditorium in the basement of the central building, the foundation and rough interior of which had previously been completed. Walls and ceilings were finished, and the sunday school rooms, the bowling alley, a gymnasium, and other rooms in connection with the church and social work were completed.

While construction work was still in progress, the church congregation moved into the basement of the building to carry on religious services and various activities.

The congregation had since then been holding such religious services in the basement and gymnasium. Bowling alleys, social halls, and other rooms were in use. The failure of the pledgers to pay the full amount of their subscriptions prevented completion of the project.

As a contractor, how does such a scenario affect you? Clearly, the obligations of Broadway Temple to the general contractor,Kennedy Construction Company were unaffected by the Temple's fund raising failure. Kennedy was entitled to be paid in full for the portion of the work completed. Kennedy was also entitled, because of Broadway Temple's breach, to the damages flowing from that breach. Those damages would certainly include his lost profit on the uncompleted portion of his contract.

CITATIONS

20 East 74th Street, Inc. v. Minskoff, 308 N.Y. 407 (1955) 3.2.3

A.D. Granger Co. v. Brown-Ketcham Iron Works, 204 N.Y. 218 (1912) 3.5.9

Acme Builders, Inc. v. Facilities Development Corp., 51 N.Y.2d 833, 433 N.Y.S.2d 749 (1980) 3.2.1

Alsens American Portland Cement Works v. Degnon Contracting Co., 222 N.Y. 34 (1917) 3.6.1

American Store Equipment & Const. Corp. v. Buffalo Municipal Housing Authority, 202 Misc. 222, 111 N.Y.S.2d 688 (Erie Co. 1952), aff'd 282 A.D. 824, 122 N.Y.S.2d 533 (4th Dept. 1953), aff'd 306 N.Y. 773 (1954) ... 3.4.2

Application of Liebhafsky, 100 A.D.2nd 764, 474 N.Y.S.2d 723 (1st Dept. 1984), aff'd Liebhafsky v. Comstruct Associates, 62 N.Y. 2d 439, 478 N.Y.S. 2d 252 (1984) 3.5.2

Arena Constr. Co., Inc. v. Town of Harrison, 71 A.D.2nd 647, 419 N.Y.S. 2d 3 (2nd Dept. 1979) 3.5.6

Arndt v. Leff, 14 Misc.2d 677, 179 N.Y.S.2d 356 (Albany C. Ct. 1958) 3.6.1

Arrow Plumbing Co. v. Dare Constr. Corp., 212 N.Y.S. 2nd 438 (Nassau Co. 1961) 3.6.1

Asbestolith Mfg. Co. v. Kerley, 129 N.Y.S. 512 (App. T. 1911) 3.6.10

Atlantic, Gulf & Pacific Co. v. Woodmere Realty Co., 156 A.D. 351, 142 N.Y.S. 953 (2nd Dept. 1913), appeal dismissed 209 N.Y. 557 (1913) 3.5.3

Barnum v. Williams, 115 A.D. 694, 102 N.Y.S. 874 (2nd Dept. 1906), aff'd 190 N.Y. 539 (1907) 3.8.8

CHAPTER 4
BREACH

4.1 GENERALLY

§ 4.1.1 Notice

When your contract is breached, what are your options? Your options remain the same whether the breaching party is an owner or your subcontractor. Upon breach, you can either rescind the contract, consider your obligations at an end and commence suing for damages, or continue the contract with the intention of seeking the damages you suffer because of the other party's breach. The breach, of course, must be substantial.

The sole fact that both parties have breached does not terminate the contract. This is especially so where each party to the contract claims that it can be enforced against the other. As a contractor with an advantageous supply contract, you are in the same position as the Glens Falls Portland Cement Co. in the following example. You assign your contract. *Hudson River Water Power Co. v. Glens Falls Portland Cement Co.*, 107 A.D. 548, 95 N.Y.S. 421 (3rd Dept. 1905), modified on reh'g 109 A.D. 919, 95 N.Y.S. 1137 (3rd Dept. 1905), aff'd 186 N.Y. 597 (1906), involved a contract by which the Hudson River Water Power Co. committed itself to supply electric power to the Glens Falls Portland Cement Co. At the time of contracting, the power company was to construct a dam and powerhouse. Pursuant to the contract, the power company would supply electricity to the cement company for a term of five years. The cement company agreed that in connection with the energy supplied by the power company, it would not use any apparatus except that of standard make and purchased from a reputable manufacturer. The motors and arc lamps were to be guaranteed by the manufacturer to have a power factor as high as any obtainable in the market. The transformers were to be guaranteed to have any efficiency as high as any obtainable. The power company would have the opportunity to test this equipment for the purpose of ascertaining that its power factor or efficiency was equal to that guaranteed. Any apparatus defective in that respect could be rejected by the power company. The cement company notified the power company that it was ready to have the power delivered. Subsequently, the cement company assigned the contract to a lighting company. The power company contended that the contract was not assignable. The lighting company requested repeatedly that the power company deliver the power. In the meantime, the lighting company had paid to the power company the first two installments under the contract.

The court found that each of the parties was trying to hold the other in default. The court noted that if both were in default, that alone would not be sufficient to justify the conclusion that the contract had been terminated. There had been no election on the part of either party or by the assignee, lighting company, to treat the contract as terminated. Except for the failure of the power company to furnish power to the lighting company upon the contention of the invalidity of the assignment, had there been any refusal to perform. This being so, there had never been an agreement between the parties to terminate the contract. Each party has concurred in at least one thing, that the contract is in full force.

The power company was not required to furnish electrical energy to a water rheostat, which had been installed by the cement company. The contract explicitly provided that energy would be furnished instead to transformers as specified in the contract. The lighting company, on the other hand, contended that it had installed motors and transformers on the premises of the cement company in accordance with and guaranteed as required by the contract. The court found that such equipment had not been so installed and guaranteed.

Until the equipment for receiving electrical energy has been installed on the premises of the cement company as provided for by the contract and guaranteed by the manufacturers, as required by the contract, the lighting company, as assignee, is in no position to demand specific performance of the contract. The court found that the lighting company was under no obligation to make the two installment payments that it did make. These payments, having been made under a mistake of fact, were properly recovered by the lighting company. However, the lighting company was not shown to be entitled to any further relief from the power company. Since the power company could not deliver electrical power in accordance with the terms of the contract, it could not recover from the cement company for breach of contract.

If you are injured by a breach of the owner and you continue the contract, what are the consequences? Your right to terminate the contract is waived. In *Emigrant Industrial Sav. Bank v. Willow Builders, Inc.*, 290 N.Y. 133 (1943), Willow entered into a contract to purchase from Emigrant an eleven story apartment building. Emigrant took a purchase money mortgage and made a building loan agreement to advance monies to finance alterations to the building. The contract provided that should the alterations not be prosecuted with reasonable diligence, Emigrant would have the right to demand all monies secured by its mortgages. The bank sought to foreclose on both mortgages, arguing lack of diligence. Although strikes did occur, the court found that they were not caused by Willow. Therefore, Willow was excused pursuant to a provision of the contract.

The court found no evidence that Willow had failed to carry on the work diligently and continuously.

The bank then argued the uncontested fact that Willow had assigned certain monies due under the contract in contravention of another contractual provision. The issue, explained the court, is whether the bank, by its silence has precluded itself from electing to foreclose the mortgages because of Willow's conceded failure to carry out the terms of the contract. It should be noted that the bank was aware that assignments had been made.

When a contract is broken in the course of performance, the injured party (here, the bank) had a choice of continuing the contract or refusing to go on. Since the bank chose to go on, it lost the right to terminate the contract because of the default. Since the bank failed to object to the assignments and in fact held discussions with Willow regarding the possibility of increasing the building loan, it is evident that the bank considered the contract to be in effect. The bank cannot now elect to foreclose its mortgage for a default, which it chose to disregard as a grounds for termination of the contract.

Remember, if an owner breaches its contract in a substantial way, the contractor need not regard the contract as terminated. The contractor may continue to perform. By performing, he or she reserves his or her right to recover any damages he or she has suffered by the breach. However, in order to recover later for the breach, he or she must perform in all essential respects. *Ferguson Contracting Co. v. State*, 202 A.D. 27, 195 N.Y.S. 901 (3rd Dept. 1922), aff'd 237 N.Y. 186 (1923), involved a contract for a portion of a barge canal.

The barge canal was a project of enormous magnitude. It was understood that certain aspects of the undertaking might not be foreseen. There was a likelihood of changes and additions to the work as it progressed. Consequently, a provision was inserted in the contract that the State reserved a right to make additions and changes in the plans and specifications. Ferguson accepted six prior changes. When the seventh change order was proposed, Ferguson refused to proceed with the work, claiming that the seventh change order constituted a radical departure from the plans and specifications and was outside the scope of the work as originally contemplated. Ferguson argued that the change constituted a fundamental alteration of the contract amounting to a breach by the State. He refused to continue work under the contract. The State alerted Ferguson that it would cancel the contract and, in fact, did cancel the contract. The remaining work was readvertised and relet. A new contract was entered into for completion of the work.

Ferguson's claim alleged that, by requiring the extra work in the seventh change order, the State breached the contract and relieved Ferguson from performance.

The court found that the State had the right to require Ferguson to do the extra work in the change order. The work was of the same general character as that specified in the contract. In fact, the materials required had unit prices specified in the contract. The principal changes were the substitute of a concrete lining in place of puddled lining; concrete docking in place of wooden docking; and additional embankment construction. The court found that these changes did not work a substantial change in either character or cost of work. This change could be made without invalidating the contract.

Although Ferguson alleged a breach by the State he did not continue his performance of uncontested aspects of the contract.

The court found that by refusing to perform the work in the seventh change order at the unit prices specified, Ferguson broke the contract. The State was justified in reletting the contract.

§ 4.1.2 Character of Breach

You can rescind a contract where there is substantial nonperformance or breach by the owner. The failure of an owner to make periodic payments constitutes a breach, entitling you to regard the contract as terminated. However, the contract cannot be regarded as terminated should the parties continue to act as if it were in effect. In *Taylor v. Goelet*, 208 N.Y. 253 (1913), a building was to be completed within a reasonable time. Work began and proceeded. The court noted that where a contract fixes a time within which a contract is to be performed and performance within that time is waived, neither party can then rescind the contract for delay without giving notice to the other that performance within a reasonable time is now required. The owner, Taylor, attempted to rescind because of a prior delay. The court held that Taylor could not so act. By Taylor's failure to act at once he cannot induce his contractor to continue performance under belief that prompt performance was waived. It is not that Taylor is without remedy. Although time as an essential element had been waived, it can be restored by a reasonable notice demanding performance. That notice should state that the contract will be rescinded if the contract is not performed within a reasonable and specified time.

In *Paterno & Sons, Inc. v. New Windsor*, 43 A.D.2d 863, 351 N.Y.S.2d 445 (2nd Dept. 1972), the town of New Windsor had contract with Paterno for the construction of a sewer trunk system. The contract had provided for periodic payments as the work progressed. The town failed to make those periodic payments. Paterno notified the town that it was in breach and that the contract was terminated. The court agreed that the town's failure to make periodic payments due under that contract constituted a breach. That breach entitled Paterno to regard the contract as terminated. Paterno can seek recovery of the balance due under the contract. The court also noted that it is well-settled law that Paterno also had the right to maintain an action for the value of the work performed and materials supplied as well as for the damages flowing from the failure of the town to permit him to complete the contract.

§ 4.1.3 Failure of Consideration

We have discussed failure of consideration earlier, at § 3.4.1. You will recall that where the very consideration upon which a contract is based, that item about which the parties bargained ceases to exist; then the contract may be rescinded.

§ 4.1.4 Inability to Perform

A contract may be rescinded because of the inability of one of the contracting parties to perform it. The contract in *Callanan v. Powers*, 199 N.Y. 286 (1910), was for reconstruction of a steam railroad and its conversion to an electric railroad. There were contracts for furnishing power for the electric railroad and railroad cars, for the acquisition of additional property, and to extend the railroad. Pursuant to these agreements and by purchase of railroad stock, the contractors obtained absolute control of the railroad corporation and possession of various bonds issued by the railroad. In addition, the contractors caused the railroad to enter into a contract with an electric light and power company owned and controlled by themselves for their own profit and to the detriment of the railroad.

The court found a nearly absolute failure on the part of the contractors to electrify the railroad. The evidence supported the finding that rescission of this contract is appropriate where there is a failure to transform this steam railroad into an electric one.

§ 4.1.5 Renunciation

What does your attorney mean when he or she speaks to you about renunciation of the contract? In what circumstances should you, the contractor, consider renouncing a contract?

There are three related concepts that must be considered: renunciation, repudiation, and rescission. Renunciation is the most general term. In renunciation, a person acts to abandon a right without transfering that right to another. Repudiation is more particularly used to indicate the renunciation of a contractual right. Finally, rescission is the act taken by the other contracting party, subsequent to the repudiation by the first party, which unmakes the contract. Once one party has repudiated the contract, the other party can refuse to be further bound by the contract, thereby rescinding the contract. Repudiation by one authorizes the other party to treat the contract as terminated.

Before you should consider rescinding a contract, you must be absolutely clear that the facts show the other party has repudiated. If the circumstances are not clear, a court may find that your acts, instead of rescinding a previously repudiated contract, in fact constitute a repudiation and breach. You would be liable for that breach.

For example, *McGowan & Connolly Co. v. Kenny-Moran Co.*, 207 A.D. 617, 202 N.Y.S. 513 (1st Dept. 1924), involved an attempted rescission by a construction contractor. McGowan claimed a right to breach its contract. It alleged a right to abrogate the contract because it had been delayed by the failure of the architects to approve the plans. This alledged repudiation would be implied to the owner, Kenny-

Moran, since the architect was acting for the owner. However, the contract contained no fixed time by which the plans were to be approved by the architects. McGowan stated it would breach the contract unless given additional compensation. The court found that for McGowan to rely on the failure to approve the plans, it should have given notice to the architects affording them a reasonable time to approve the plans. In the absence of such notice that McGowan was prejudiced, there was no right of rescission.

The court noted that McGowan's real grievance was not the delay in the approval of the plans, but an increase in labor and material. McGowan had repeatedly submitted to Kenny-Moran copies of drawings for approval by the architects without complaint of delay. The court found no question of delay until the owner, Kenny-Moran, requested that McGowan proceed with the work.

What are the legal considerations for McGowan's demand for extra compensation? Is there new consideration for the additional compensation, or has the original contract terminated by consent, and a new one executed in its place? Any change in an existing contract, such as a modification and increase in compensation, must have new consideration. In such a case the contract is continued. In the alternative, the existing contract may be terminated by consent of both parties, and a new one executed in its place.

McGowan did not demonstrate that there was additional consideration to form a basis for Kenny-Moran's paying additional compensation. Neither did McGowan prove that the existing contract was terminated by consent and a new contract substituted.

§ 4.1.6 Rescission by Building Contractor

The clearest situation in which rescission occurs is at the bidding stage. *Syracuse v. Sarkisian Bros., Inc.*, 87 A.D.2d 984, 451 N.Y.S.2d 945 (4th Dept. 1982), aff'd 57 N.Y.2d 618, 454 N.Y.S.2d 71 (1982), demonstrates that at the bidding stage classic rescission is possible. Rescission places both parties in the position, which existed immediately prior to even the thought of a contract. The City of Syracuse had solicited bids for the construction of a junior high school. Sarkisian had prepared a preliminary bid proposal in accordance with the plans and specifications. Just prior to bid, Sarkisian determined it could reduce its bid by $21,300. In transposing the reduction from a worksheet to the bid proposal, the sum of $213,000 was inadvertently deducted. On the day the bids were submitted and opened, Sarkisian discovered its error and immediately notified the city. Sarkisian requested that the bid be withdrawn. Syracuse requested Sarkisian's worksheets, which were promptly forwarded. Three weeks after the bid submission and notification of the error, Syracuse awarded the contract to Sarkisian based on its erroneous bid.

Sarkisian declined to perform and Syracuse readvertised the job. The contract was awarded in an amount closely approximating the bid, which Sarkisian had intended to submit. Syracuse brought an action seeking money damages and forfeiture of Sarkisian's bid bond. Sarkisian asked the court to rescind the contract.

The court found that the clerical error made by Sarkisian was excusable. The error was material and of such an amount to make enforcement of the contract unconscionable. The error could be objectively determined by comparing the work sheets with the bid proposal. The court noted the criteria for rescission of a contract. A bid is a binding offer to make a contract. The bid may be withdrawn in the case of a mistake by the bidder, where the mistake is known to the other party to the transaction and (1) the bid is of such consequence that enforcement would be unconscionable, (2) the mistake is material, (3) the mistake occurred despite the exercise of ordinary care by the bidder, and (4) it is possible to place the other party in status quo. The court found that all these criteria were met. Sarkisian's bid was rescinded. The court noted that Syracuse was not damaged since it received prompt notice of the error and could have awarded the contract to the second lowest bidder. There was never a meeting of the minds, which would have established a contract since Sarkisian never submitted its real bid, but instead, an erroneous one.

Since the bid is rescinded, Sarkisian's bid bond is canceled. Sarkisian had no legal obligation on the bond.

In *Cranford Co. v. New York*, 150 A.D. 195, 134 N.Y.S. 839 (1st Dept. 1912), aff'd 211 N.Y. 534 (1914), Cranford was to sink two exploratory shafts on either side of the Hudson River to determine the elevation of solid rock. In addition, the suitability of the rock for containing the Catskill Aqueduct tunnel was to be determined. The contract provided that not more than once a month the city engineer would make an estimate of the work performed and materials furnished and that the city would pay Cranford 95 percent of each estimate. No time was specified within which the city should pay the 95 percent. Three monthly requisitions were pending with the Comptroller when Cranford made a demand that unless they were paid within three days, it would discontinue the work.

There is authority for the proposition that the failure of a party to a contract to pay installments when due is a breach of the contract. The failure to pay is, in effect, a repudiation. It absolves the other party from all obligations to perform on its part while the breach continues.

Was there such a breach by the city? Since no time for payment was stated, a reasonable time is implied. The court noted that what might be construed as inexcusable delay on the part of an individual, partnership, or corporation might not be such in the case of a municipal corporation. The contractor who deals with a municipality should not expect the same promptness of payment. The court noted that when Cranford preemptorially demanded payment of all three monthly estimates, the question of whether Cranford had the right to rescind the contract must be tested with reference to the most recent estimate. Since the delay of that estimate was only 24 days, the delay was not unreasonable. Cranford had no right to rescind since the city had not breached. Cranford's suit was dismissed.

With the dismissal of Cranford's suit the court found it unnecessary to discuss the measure of damages that would have been applicable if Cranford had been justifiable in its decision. Cranford sued on the basis of the reasonable value of the work done by it, less the sums it had been paid at the contract rate. The testimony had indicated that Cranford had underbid the job. Its claim for reasonable value was in excess of $200,000 although the three unpaid estimates barely exceeded $25,000. In the court's opinion, Cranford would not, in any event, recover more than the contract price of the work actually performed. Cranford did not contend that the city was guilty of wanton or willful breach. Under these circumstances, the contract price would have limited the damages.

The questions of rescission and the measure of damages for rescission again discussed in the case of *Ryan v. Rodgers & Hagerty, Inc.*, 197 A.D. 662, 189 N.Y.S. 269 (1st Dept. 1921), Rodgers had a contract to build a section of a subway. He subcontracted to Ryan the handling, laying, and rodding of tunnel ducts. Ryan was to be paid 2 3/4 cents per duct foot. Hagerty was to withhold 1/2 cent per duct foot as retainage. Some time after the signing of the subcontract, Ryan approached Hagerty and advised him that he would need the contract monies for payroll. Hagerty ordered his engineer to measure the work. Hagerty paid Ryan, but withheld the 1/2 cent per foot. Over the course of the project, payments were made upon request by Ryan in order to meet his payroll. Once all tunnel ducts were constructed, a small balance, less than the 1/2 cent per foot, remained. Ryan claimed it was owed its contract balance and profit on other allegedly uncompleted work. The court found that all of Ryan's work had been completed. The court noted that even if Ryan had been correct and there was remaining work, the failure of Hagerty to pay him the amount claimed to be due left Ryan with one of two options, Ryan could either rescind the contract and recover what was due under the contract, or he could proceed with the contract and sue for the amount due. The court found that Ryan could not rescind the contract and recover prospective profits.

This statement by the *Ryan* court is a curiosity. The court found that Ryan had completed its contract, therefore his profits would have been included in the contract balance. The general rule is that when an owner breaches, the contractor is entitled to his or her damages. Those damages may be either the reasonable value of the work performed or, under the contract, the contract value of the work performed and the profit on the unperformed work. Certainly, the contract value is a convincing indication of the reasonable value of the work. The court in *Ryan* is dealing with a situation involving both a completed contract, where lost profits would not enter into the equation, as well as a contract that was underbid to begin with. In its attempt to do justice, the court has misstated the measure of damages. The curious aspect of *Ryan* is that it has never been overruled and, in fact, has been cited by a Wyoming court. The fact remains, upon breach, the party injured by the breach can either seek the reasonable value of its work or the contract value of its work and lost profit on the portion of the contract uncompleted.

§ 4.1.7 Rescission by Building Owner

Should the contractor default or repudiate the contract, the owner has the right to rescind as well. The breach, of course, must be material and not a slight or technical breach. *Fisher v. Goodrich*, 61 A.D. 534, 78 N.Y.S. 38 (2nd Dept. 1901), illustrates this proposition. Fisher contracted to provide a complete windmill. His proposal to Adelaide Goodrich contained a list of the materials to be provided. Fisher furnished the pump, the tower, and the other materials. The tower, however, was insufficient to support the tank. It was clear that it was not safe.

The court found that the structural instability of the tower was of a material nature. Miss Goodrich had the right to terminate the contract and to require Fisher to remove the windmill. In explanation, the court continued that had Miss Goodrich agreed to a substitution of a new tower and Fisher procured one, then she could not prevent the tower's erection. Such a prevention would have been a breach of the contract on her part and would have entitled Fisher to recover his damages. Rescission by Goodrich was proper.

In *Reddy v. Barney-Ahlers Const. Corp.*, 188 N.Y.S. 176 (App.T. 1921), Reddy had a subcontract with Barney-Ahlers to remove dirt and rock for the cellar of a factory to be constructed by Barney-Ahlers. Barney-Ahlers subsequently rescinded the subcontract, basing that rescission on a delay in completion by Reddy. Reddy's contention was that the cancellation was unjustified. Delay in completion was caused by the failure of Barney-Ahlers to furnish proper plans and levels from which to calculate the depth of the excavation. The evidence supported Reddy's contention that the rescission was unjustified. Notwithstanding, Barney-Ahlers's claim that a smaller quantity of work was performed by Reddy, the court found Reddy's testimony persuasive. Barney-Ahlers had difficulty in accurately checking the actual work done during the life of the subcontract. This difficulty would account for any discrepancies in the quantity calculations. Since Reddy was wrongfully discharged he can recover at the subcontract rate for work done under his subcontract before cancellation and subsequently utilized by Barney-Ahlers.

If time is of the essence in your contract, the contract may have a provision that permits termination upon an architect's certificate that you have failed to prosecute the work diligently. If that contract makes the architect the sole arbiter between the parties to the contract, then the architect's estimate as to quantities is conclusive. The contract in *Midtown Contracting Co. v. Goldsticker*, 165 A.D. 264, 150 N.Y.S. 809 (1st Dept. 1914), for alterations to a building contained a provision, which stipulated that should the contractor fail to supply a sufficient number of properly skilled workmen or materials of proper quantity, and that should such failure be certified by the architect, then the owner could, after three-days-written notice, provide such labor or materials. The provision continued that the architect could also certify that refusal, neglect, or failure of the contractor was sufficient grounds for the termination of the contract. That clause was interpreted by the court to mean that a three-day notice was necessary to supply labor or materials to be deducted from the contract price, but that no notice was necessary to the contractor, upon certification by the architect for termination.

§ 4.1.8 Delay

Will delay in completion of the contract be deemed repudiation, which allows the other contracting party to rescind? Where time is of the essence, the answer is yes. Where time is not of the essence, however, delay in completion will not be deemed material, and thus will not give rise to the right of rescission.

The parties to the contract need not wait for prompt performance. If time is not of the essence in the contract, it may be made of the essence by notification to the other contracting party that time is now the essence and performance must be completed by a specified date that allows a reasonable time for notice and completion.

What is appropriate notification? In *Keilbert Const. Co. v. Frey*, 174 A.D. 513, 161 N.Y.S. 156 (1st Dept. 1916), Keilbert Construction was selling certain properties to Frey. Dates were specified for securing a permanent mortgage, but time was not made of the essence. Subsequently, Frey contended that time had been made of the essence. Frey's attorney had written a letter outlining certain terms of the contract and his interpretation of those terms. The court found that this was not enough to make time of the essence. The letter was a mere statement of the construction of the original agreement as interpreted by the attorneys for Frey. Such a letter was not notice that unless Keilbert performed on or before a certain time, then the agreement would be terminated. The court noted that even if it were deemed sufficient notice, the default was waived by a subsequent acceptance of the mortgage. The moral of the story is that if the contract is to be terminated, the words of termination must be absolutely clear.

§ 4.1.9 Damages

The damages we are discussing are liquidated damages. Almost without exception, every contract you sign will have a liquidated damages provision. That provision is tied in to the contract completion date. If the contract is not completed by that date, the owner is entitled to a set amount per day in damages. The *per diem* amount must bear a reasonable relationship to the actual costs for the delay. However, the courts recognize that since delay damages are difficult to ascertain, a reasonable approximation will be enforceable. The *per diem* amount must not be a penalty.

In *Triad, Inc.*, 89-2 B.C.A. 21,661 (V.A.B.C.A. 1989), the owner's proof sufficiently supported the liquidated damages amount. Triad contracted to replace windows at a number of buildings at the Veterans Administration Medical Center in Fayetteville, North Carolina. It was uncontested that the contract was completed some time after the extended contract completion date. The court noted that liquidated damages are appropriate from the extended contract completion date to the date of substantial performance.

Just when did substantial performance occur? The Board of Contract Appeals found that substantial performance occurred more than seven months earlier than the date chosen by the Veterans Administration. It based this determination on a number of reasons. The date chosen by the Veterans Administration was the date when final trim was placed on the windows. The Board found that the government had failed to prove any adverse effect due to the absence of trim. The Board particularly noted that there was no complaint by the government that it was waiting for the trim, nor notification to the contractor that the windows were not performing their intended function and that liquidated damages were considered to be running. Consequently, the Board found that substantial completion occurred upon installation of the windows. The absence of trim was not a material deficiency that would prevent a determination that substantial completion had occurred.

Triad also argued that the *per diem* amount of liquidated damages was excessive and did not represent a fair approximation of the expected damages for delay. In addition, Triad argued that the liquidated amount should be prorated.

The first question to be considered is whether or not the liquidated amounts are appropriate. If they are not, they may be construed as a penalty and stricken. However, if the damages do properly approximate actual damages that are difficult to quantify, they will be allowed.

The court acknowledged two approaches regarding the application of liquidated damages. Under the first line of cases, where the parties make no provision for the apportionment of liquidated damages among various aspects of the work, the court should not attempt to modify the agreement. The liquidated damages are allowed in the full amount. The Board in *Triad* noted that this appeared to be a minority position and, consequently, attempts to reconcile this position with the second approach.

The second approach allows liquidated damages to be prorated, where the work is severable and identifiable. The liquidated damages must be tied directly to the severable portions of the work. If liquidated damages are based on expected costs to rent equivalent space in, for example, a contract to construct a number of separate buildings, then the work can be severed and the liquidated damages correspondingly severed.

The Board noted, however, that where there is no direct relationship between the establishment of liquidated damages and the value of the severable portions of the work, then liquidated damages will be enforced in full. The testimony in *Triad* was that the liquidated damages were based on an estimated administrative cost, plus a minor amount for energy savings. Since the administrative costs would not necessarily diminish as the buildings were completed, the liquidated damages will be enforced in full. The Board found no rational basis for prorating liquidated damages.

As for the apportionment question, how should that best be handled? By the parties themselves. If there is concern that a court will not apportion the liquidated damages, then the parties, in order to avoid this uncertainty, should insure that the liquidated damages provision allows for apportionment.

While a contractual provision for liquidated damages is enforceable, where that provision is construed by the court to be a punishment or penalty, it will be considered invalid and nonenforceable.

How does a court evaluate whether a liquidated damages provision is in actuality a penalty? In *Sides Constr. Co. v. City of Scott City*, 581 S.W.2d 443 (Missouri C.A.

1979), the contract was for a public swimming pool. Liquidated damages were to be assessed at the rate of $50.00 per day, and time was of the essence. Provisions for a fixed *per diem* payment for delay in performance of construction contracts are usually construed as stipulations for liquidated damages and not as penalties. This is particularly true where actual damages are uncertain or difficult to ascertain and where the stated *per diem* amount is a reasonable estimate of probable damages or reasonably proportionate to actual damages. The courts must consider the nature of the contract, the public detriment, and the difficulty in approximating the damages.

The court noted that the precise amount of damages was difficult to ascertain. This is precisely why a liquidated damages provision is beneficial to both sides in a construction project. It protects the owner from a laborious recitation of item-by-item damage. It protect the contractor from a potential lump sum claim for substantial sums of money in the event of a breach.

The court performed a mathematical analysis. It noted that the contract for the swimming pool was $274,000. Liquidated damages were $8,300, which was only 3.029 percent of the total contract amount. The court found that the amount of liquidated damages was not oppressive and was not greatly disproportionate to the total amount of the contract. Since the court did not find the liquidated damages amount to be a penalty, it upheld the provision.

In preparing the *per diem* amount of liquidated damages, what considerations should an owner include? Let us consider a contract for the construction of a building. Certainly, the costs of rental space. In addition, if the owner is operating in an old or inefficient building, it may consider its additional maintenance and utility costs. Does the owner incur maintenance costs of the as-yet uncompleted building? If it does, those should be considered as well. Of course, extended supervision and engineering costs should be included as well as any labor and material cost increases.

When does substantial completion occur? *Dillon Constr., Inc.*, 81-2 B.C.A. 15,416 (Eng. B.C.A. 1981), involved the construction of two parking lots and ancillary lighting. With respect to the first lot, the Board found that, since the lot was in use, as intended, by the general public, the fact that two street lights had not been installed was not material. There was sufficient lighting from adjacent sources and from street lights.

With respect to the second lot, substantial completion occurred once the government took possession of the lot. The Board found that the lack of lighting as of that date did not dictate a contrary result. The attainment of substantial completion ended the accrual of liquidated damages.

Do liquidated damages accrue on weekends and holidays? To avoid this question, the liquidated damages provision should specify calendar days, inclusive of weekends and holidays, rather than working days.

It is important to remember that for an owner to collect liquidated damages he or she must have performed the contract on his or her part. Where the owner has breached, or the contractor's conduct in not completing is reasonable under the circumstances, the owner cannot recover his damages. *Sutton Constr. Co.*, 68-2 B.C.A. 7250 (A.S.B.C.A. 1968), involved the installation of an air ventilating unit.

Once the government had learned that Sutton planned to use a certain manufacturer, the project officer asked for and received permission to consult directly with the manufacturer. After consulting with the manufacturer, the government ultimately rejected the contemplated ventilating unit because it failed to meet the contract specifications. However, the government did not promptly notify Sutton. After notification, Sutton submitted the drawings of a unit from another manufacturer. The unit was approved and delivered six days after the time for completion of the contract had expired. The Board found that both parties had contributed to the delay. It was impossible for the contract to be completed on time. Since the delay was caused by both parties, the Board did not attempt to apportion responsibility, but simply found that the liquidated damages provision was annulled.

Clearly, liquidated damages are applicable for late completion of a contract. But, suppose the contract is not completed. Suppose, instead, that the contractor abandons his or her work. *Village of Canton v. Globe Indemnity Co.*, 201 A.D. 820, 195 N.Y.S. 445 (3rd Dept. 1922), involved the construction of a waterworks system. Subsequent to the contract completion date, the contractor abandoned the contract. The court noted that not only was the contractor being paid its progress payments, but was also receiving advances from the village. It also noted that the parties had agreed on a change of location in the pipeline. The court declined to enforce the liquidated damages provision. The court held that the provision for liquidated damages applies only to delays and not to abandoned performance. The court noted the anomaly that an owner should be allowed liquidated damages for delay in the completion of an abandoned contract, which the owner itself is completing. Although the village may have completed the work diligently, there is a constant temptation for an owner not to hasten the work to completion. A liquidated damage provision is read to apply only to work performed by the contractor, not by the owner.

Town of North Hempstead v. Sea Crest Constr. Corp., 119 A.D.2d 744, 501 N.Y.S.2d 156 (2nd Dept. 1986), involved the construction of a solid waste treatment plant. Sea Crest abandoned the project, and the town sued, seeking actual damages. In a curious twist, Sea Crest attempted to use the liquidated damages provision as a shield, arguing that the provision precluded the town from seeking actual damages. Not so, said the court. The provision is not so clear and unambiguous as Sea Crest would like the court to believe. The liquidated damages provision applied to situations in which damages were sustained due to delays in Sea Crest's full performance of the contract. The provision did not limit the amount of actual damages the town may recover upon Sea Crest's abandonment of the contract. The court held for the town.

The importance of accurate, inclusive contract drafting is reaffirmed by New York's highest court in *City of Elmira v. Larry Walter, Inc.*, 76 N.Y.2d 912, 563 N.Y.S.2d 45 (1990). The breached contract was for the construction of a municipal parking garage. Disputes arose during construction, and Larry Walter discontinued work on the project. The City of Elmira was forced to retain another contractor to finish the contract. At the trial level, Larry Walter was found to have breached, and the City was awarded actual as well as liquidated damages. At the intermediary appeal level, the liquidated damages portion of the award was stricken. The state's highest court subsequently determined that the city was not entitled to an award for liquidated damages.

The liquidated damages provision on which the City relied was an attempt to anticipate and provide for the specific possibility that Larry Walter's completion of the project might be delayed beyond the contract completion date. The court noted that the provision does not contain clear and unambiguous language as it should, indicating that it was also intended to apply to Larry Walter's outright abandonment of the project. The court found abandonment to be an entirely separate eventuality. The reasonable interpretation of the provision was that liability for liquidated damages would not accrue until the contractor had fulfilled the contract. The provision is not available in a case such as this involving a complete repudiation of the contract. The City had argued that this interpretation of the provision would encourage contractors to abandon their contracts as a means of avoiding the agreed-upon liquidated damages. The court discounted this contention and opined that those owners who wish to protect themselves should negotiate an express contractual provision for liquidated damages even in the event of an abandonment.

Your contract contains a liquidated damages provision. Its application would appear straightforward. It ain't necessarily so. Your lawyer can best advise you as to its applicability. Is this a true approximation of the actual damages, or is it a penalty? Can the owner seek actual damages in addition to liquidated damages, or does the provision preclude this? Finally, as in any contractual provision, the owner can waive a liquidated damages provision. Has the owner, as in *Triad*, failed to alert you that it intends to enforce the provision after a delay occurred? Or, has waiver occurred because, as in *Sutton*, the owner has allowed you to complete your work after the contract completion date has passed? Or, will the courts find waiver because the owner has continued to make progress payments without a deduction for liquidated damages? These are all issues that should be discussed with your attorney. The liquidated damage provision may simplify proof of damage; however, its applicability is not always that simple.

4.2 CONTRACT RESCINDED

§ 4.2.1 Failure of Consideration

The issue here is restitution. Where consideration fails, the very object for which the parties bargained, how are the contracting parties restored to their respective positions? The classic case is *Tompkins v. Dudley*, 25 N.Y. 272 (1862). The trustees of the school district sought the money advanced by them for the construction of a schoolhouse. When the building was substantially completed, but before it was turned over to the school district, it burned to the ground. Since the building had never been turned over to the district, the contractor had not become entitled to the contract amount. The contract created a liability and the contractor incurred a duty that he guaranteed he would perform. He alleged that the destruction of the building was an inevitable accident without any fault on his part. The contractor could have taken comprehensive builder's liability insurance. In the alternative, the contract could have allocated the liability for damage by fire. However, where the contract is silent, that liability is with the contractor. The contractor has no defense to nonperformance; therefore, the amount advanced by the district must be returned.

The concept of restitution was also considered in *First Church of Christ Scientist v. Schreck*, 70 Misc. 645, 127 N.Y.S. 174 (Eire Co. 1911). Involved in that case was a fund for purchase and placement of an organ in a new church building. Members of a Sunday school contributed to the fund. The secretary and treasurer of the school, Clara Schreck, reported to the annual church meeting with respect to the amounts in the organ fund. For 15 years, she maintained the fund. At the end of 15 years she withdrew from the church along with the monies on hand. When the church demanded that she pay over the fund, she refused to do so. The court naturally found for the church and ordered that the funds under Ms. Schreck's control, with accumulated interest, be paid to the church. The court does instruct, however, that the church is to use the fund for the specific purpose of providing a new organ in the new church when erected. The church has no right to apply it to any other purpose. If that purpose is abandoned, the donors have the right, if they so elect, to demand restitution of the fund to them by the church.

§ 4.2.2 Part Performance

Are you entitled to restitution if you partially perform your contract knowing that the owner has repudiated? Where a contractor has knowledge that the owner has repudiated, courts will not permit partial performance to increase the total amount that would be awarded in restitution.

§ 4.2.3 Definitely Apportioned Exchange

What is meant by a definitely apportioned exchange? Consider *Tompkins v. Dudley*, at § 4.2.1. Tompkins was entitled to restitution since the contractor had not completed the school and refused to perform further. Suppose the contract had been for the construction of two schools, and the amount to be paid was allocated between the two schools. One school is completed and turned over to the school district. The second school is almost completed and burns to the ground. Is the school district entitled to restitution of the amounts advanced for the construction of both schools? Of course not. Restitution could not be sought for the portion of the contract that was paid for and for which the district received its benefit; i.e., the first school, since the contractor's performance there had been complete.

Consider another hypothetical. Suppose the contract again is for two schools and the amounts to be paid are allocated between the buildings. Both schools are completed but not turned over to the district. Both burn to the ground. Even though there is a definitely apportioned exchange, neither portion has been completed in full. The district could seek restitution for the full amount of the contract.

§ 4.2.4 Full performance

Again, considering the facts of *Tompkins v. Dudley*, § 4.2.1, suppose that the school district had paid completely for the school. Would the contractor's failure to perform entitle the school district to seek the full amount paid in restitution? Yes, it would.

§ 4.2.5 Rescission

A basis to rescind is the restoration by the rescinding party of any benefit received from the other contracting party. Since rescission restores the parties to their respective positions before the contract was made, all benefits of the contract must be returned. *Naimoli v. Massa*, 81 Misc.2d 431, 366 N.Y.S.2d 573, (C.Ct. Geneva 1975) involved a claim for restitution after the rescission of a contract for the purchase of land. Just prior to the contract signing, city health officials, by letter, advised that the premises were unfit for human habitation and must be vacated or repaired. Neither the seller nor the purchaser were aware of the contents of this letter when the contract was signed. Immediately upon signing the contract, Naimoli entered into possession, collected rents and commenced cleaning, decorating, and performing plumbing repair work. Naimoli also spent funds on publishing rental advertisements and paying for gas and electric service. Once the contents of the city's letter became known, both parties entered into negotiations to rescind the contract. The rescission of the contract stated that it was annulled and canceled, that the sellers would return all tax adjustments to the purchasers, and that the purchasers would return all rents to the sellers.

Naimoli sought restitution of all sums spent while in possession of the subject premises. Restitution is the finding that one person is to be accountable to another on the grounds that he or she would unjustly benefit or that the other would unjustly suffer loss. Unjust enrichment need not be unlawful or wrongful enrichment. A person is enriched if he or she receives a benefit from another that in any way adds to his or her security or advantage, or where the other saves him or her from expense. The general rule is that where a person has received benefit from another, he or she is liable to pay in restitution only if the circumstances of the receipt and retention of that benefit as between the two persons indicate that it is unjust for him or her to retain it.

The court found that Naimoli is entitled to restitution with respect to the plumbing bills and the repair of the premises. Massa received the whole benefit from this expense. As to the items of expense for painting and cleaning supplies, newspaper advertising, and gas and electric bills, Naimoli is not entitled to restitution. There was no proven benefit to Massa for these expenditures.

§ 4.2.6 Breach of contract

Whether there is a rescission or not, upon nonperformance or substantial breach by one party, the other may demand restitution. The principal is illustrated both in *Naimoli* and in *Syracuse v. Sarkisian Bros., Inc.*, at § 4.1.6.

§ 4.2.7 Illegal Contracts

Here, restitution does not apply. The law will not sully its hands in the mire of illegal agreements. The courts will leave parties to an illegal agreement where it finds them.

However, where a penalty is imposed upon one of the parties by statute, and the contract is otherwise unobjectionable, the other party, not at fault, can recover against the penalized party for any benefit the penalized party received.

Arnot v. Pittston & Elmira Coal Co., 68 N.Y. 558 (1877), involved a contract for the sale of coal. Arnot, the assignee of the Butler Colliery Company sought to recover the price of approximately 2,700 tons of coal sold to Pittston. Pittston contracted with Butler to control the shipment and supply of coal for the Elmira market in order to maintain an unnaturally high price and to prevent competition. Butler was aware of these intentions at the time it made its contract. Pittston did not agree to take the whole product of Butler's mines. It took a limited amount. By contract, Butler would not sell coal north of the Pennsylvania and New York state line. This effectively restricted the supply of coal to the Elmira market. The court noted that this attempt to create a monopoly was adverse to the interests of the public, contrary to all public policy, and was illegal. The purpose of Pittston was against public policy, and Butler knew it. The question is whether Butler, delivering goods under such a contract, can recover for the price. It could not. There is no evidence that Butler rescinded the agreement on the grounds of its illegality. Rescission does not give Butler any better right to recover for its past performance.

§ 4.2.8 Void Contracts

Restitution does apply here. Since a void contract is considered a nullity, any consideration transfered or monies paid may be recovered.

In *Dare Constr. Corp. v. Gilman*, 6 Misc.2d 160, 157 N.Y.S.2d 984 (Nassau Co. 1956), Dare sought the balance due on a contract in connection with the Gilmans' purchase of certain land with a house to be erected thereon by Dare. The Gilmans sought $2,000 representing an amount paid by them during the course of negotiations.

The court held that the receipt of the $2,000 is insufficient to be a contract in itself. On its face, it disclosed the absence of a meeting of the minds, that essential element of a contract. The receipt indicated that the $2,000 represented payment on account of the purchase price of the property. However, the receipt also stated that the balance of the purchase price, $108,000, would be paid in installments as may be agreed upon and that were to be specifically reflected in a formal agreement. The agreement was to be not only for the sale of the land, but the erection and construction of a finished dwelling. The receipt contemplated the execution of a formal contract at a subsequent time. It clearly demonstrated that at the time the receipt was given, the parties had not agreed upon the type of dwelling house, the manner in which it was to be constructed, or how payment was to be made. At the time the receipt was given, there were no plans or specifications for the house to be erected on the real property. The receipt expressly stated that the formal contract was to be executed by Dare and the Gilmans after the plans and specifications in the process of being prepared were approved as to form and substance. The receipt further stated that if a formal contract were signed, the $2,000 would be applied on account.

The parties were never able to agree on a formal contract. It follows that the Gilmans were entitled to a return of the monies they paid.

§ 4.2.9 Want of Consideration

Want of consideration indicates that a contract was neither void nor unenforceable, but simply never came into existence. Where a party to a supposed contract has paid monies or transfered benefits to another under such a false supposition, that party is entitled to restitution of the monies or benefits.

4.3 MISTAKE

§ 4.3.1 Rights of Intended Grantee

It is a general principal that where a person by mistake confers a benefit on another, that other person is liable in restitution.

One Fifty-Seven Prince Street Corp. v. Michelini, 62 N.Y.S.2d 148 (N.Y. Co. 1946), aff'd 271 A.D. 777, 66 N.Y.S.2d 406 (1st Dept. 1946), involved the purchase of certain real property. Here, the grantor, Prince Street Corp., sought to recover from Michelini the amount of the refund allowed by the Board of Estimate on an assessment against the conveyed property. The parties to the contract assumed the assessment was legal. This proved to be a mistake. Shortly after closing, the courts reversed the assessment on the ground that a proper method of computation had not been used and that the assessees had not been accorded adequate opportunity to be heard. The final assessment reduced the amount by almost half. At closing, Michelini was allowed an amount in excess of the assumed value of the assessment. There is no question that had the grantor Prince Street Corp. paid the assessment, it would have been entitled to a refund. Michelini paid nothing toward the assessment, she merely paid one-half of the amount allowed to her for the specific purpose of paying it. No allowance was made in reduction of the purchase price. The allowance to Michelini was made under a mutual mistake of law and fact for the express purpose of discharging this particular tax assessment. If Michelini was unjustly enriched, Prince Street Corp., was entitled to restitution.

4.4 FAILURE TO MAKE RESTORATION

§ 4.4.1 Requirement of Restoration

Let us start at the beginning. Rescission is most often employed with respect to a mistake in bidding, as in *Abner M. Harper, Inc. v. Newburgh*, 159 A.D. 695, 145 N.Y.S. 59 (2nd Dept. 1913). Harper was the low bidder on a street paving contract. After the bids were opened and a resolution in favor of awarding the contract to Harper had passed the City Council, Harper discovered a mistake in his bid. Harper had transposed two items. The mistake was not due to Harper's negligence.

The court considers whether rescission is appropriate where the mistake is by only one of the parties. The court first reviews the law governing *reformation* of contracts. Reformation is the principle by which the courts rectify or reform a contract ,which, because of mutual mistake or fraud, fails to reflect the real intent of the parties. The court determines that intent and applies it to the contract. Reformation of the contract only applies where there is a mutual mistake. In *Harper*, where the mistake was of one party only, for reformation to apply there must be a showing of fraud or some other inequitable conduct. If there is no such showing, a court reforming the contract might be doing right for the person who seeks reformation and yet an equal wrong to the other person, without his fault. The court found that reformation would not apply here.

Rather, the court applied the principles of rescission. The court noted that in rescission there remains no contract at all. In the eye of the law, there was no meeting of the minds. Therefore, the courts are free to rescind a contract for the mistake of one party only without a finding of fraud or inequitable conduct.

Was there a completed contract? Not only was a contract in writing contemplated, but it was required by the City of Newburgh charter. There was no proof that any contract was ever executed or even prepared for execution. The court found a legal relationship arising from the bid and the resolution of award passed by the City Council and approved by the mayor. The court noted that is was not essential that there should be a completed contract before there could be a rescission. Rescission could be used not only in connection with a contract, but with a mere legal relationship subsisting between two persons. However, noted the court, rescission involves inquiry as to whether those persons can be put back in their original positions. If that cannot be done, then the relief of rescission exists only where the clearest and strongest equity demands it. The court allowed Harper to withdraw its bid.

Now, suppose the installation is complete but does not function. *Viennese Melody, Inc. v. Majestic Refrigerator Corp.*, 279 A.D. 243, 108 N.Y.S.2d 879 (1st Dept. 1951), involved the installation of an air conditioning system in Viennese's restaurant. For approximately a year after installation Majestic endeavored to have the cooling system function. It was never successful. Viennese finally had another air conditioning firm make extensive repairs and replacements. Viennese sought to have the balance of the amount due to Majestic canceled. If Viennese retains the plant and fails to offer to return it, the court will not allow cancellation of Viennese's indebtedness. Viennese is awarded the amount of expenses it incurred in having necessary repairs and replacements made. The amount of the repairs was not excessive. All the work was directly traceable to the fault of Majestic. Certain wear and tear on the machinery seemed excessive. Would Majestic be responsible for the repairs to this machinery as well? The court noted that wear and tear on certain parts of the equipment did not result from ordinary use of the machinery, but from the many futile attempts to make it operate suitably. The court allowed Viennese the costs of its repairs to the system. However, since it retained the plant, it is responsible for the outstanding balance.

McFayden v. Nassau Chemical Cesspool Cleaning Co., 141 N.Y.S.2d 92 (Nassau Co. 1955), is a case where the law and engineering practice intersect. Nassau Chemical Cesspool had contracted with John Kollner to build a storm drain 30 feet deep on Kollner's property. Kollner made an initial down payment. From McFayden, Nassau Chemical hired a crane and operators. While proceeding with its work on the property, several unforeseen developments occurred. Sewage flowed into the excavation. The ground was not an actual deposit of unconsolidated sand, gravel, and stratified clay, but rather filled land. Nassau Chemical encountered large stumps of trees, tree trunks, and tree limbs, all of which made digging unpredictable and difficult. No adequate drainage could be obtained without piercing a hardpan layer of clay that was much deeper than contemplated. The original contract with Kollner was abandoned and a new arrangement entered into. The court found the first contract was clearly abrogated. It was succeeded by the new oral agreement. Nassau Chemical would proceed to install a proper catchment basin within a reasonable time. Kollner would pay the fair and reasonable value for the work and labor performed and materials furnished.

However, the work was not performed either within a reasonable time or in a good and workmanlike manner. Subsequently, Kollner ordered Nassau Chemical to cease the work. Nassau Chemical was attempting to sink prefabricated concrete rings into the ground. The rings, with inside diameters of 90 inches, were 4 1/2 feet wide. The concrete was 4 inches thick. In order to get the 8- foot pipe into the ground, Nassau Chemical had dug a hole 40 feet in diameter and 20 feet deep. At the conclusion of Nassau Chemical's work, there were two setups of rings not in alignment and in a condition indicating no hope of completion. Kollner hired Eugene DePasquale, an expert in the field. DePasquale removed the broken, unusable rings. He installed rings 120 inches in diameter, 6 feet wide and having a 7-inch wall thickness to a depth of 54 feet. The lower 6 feet was in sand. The hardpan clay was pierced. Therefore, the drainage basin could perform satisfactorily. The court noted that the difference in installation accounted for DePasquale's success and Nassau Chemical's failure. While Nassau Chemical sought to dig a large, open hole and set up concrete forms therein without strapping the forms together. This was unsuitable for the requirements of the job.

DePasquale's method was called a *caisson installation*. The concrete ring was laid on the ground and, with a crane, the earth was excavated from within. The first ring had a cutting edge on the underside. As the hole was dug away, the sides of the ring settled. Another ring was then placed above it and strapped to it with steel bands. The process was repeated until the installation became complete to its full and proper depth. Any materials such as stumps, rocks, boulders, and trees impeding the sinking of the caisson were cut away without enlarging the hole beyond what was required to fit snugly to the outer surface of the rings. The caisson method disturbed little of the ground beyond where the drain was located and provided an economy of excavation.

The court noted that the open pit method appeared to have been doomed to failure from the beginning. The work of Nassau Chemical was of no use to Kollner. Kollner is not required to pay for it except as to the rings that DePasquale used and that Nassau Chemical furnished. Nassau Chemical was not chargeable with work that DePasquale did for Kollner. However, Nassau was obligated to return Kollner's down payment, less the amount of the usable work performed by it.

§ 4.4.2 Necessary Requirements

What is necessarily required of you if you wish to rescind your contract? The general rule is that a party rescinding must restore or offer to restore any benefit he or she has received because of the contract. Rescission attempts to place each party to a contract in its respective status quo before the making of the contract. Therefore, restoration must be made of anything of value.

P.S. & A. Realties, Inc. v. Lodge Gate Forest, Inc., 205 Misc. 245, 127 N.Y.S.2d 315 (N.Y. Co. 1954), involved a contract for the purchase of realty. The seller had misrepresented the terms of leases for tenants and rentals. These misrepresentations were material, were intended, calculated, and likely to influence a respective purchaser and bring about a sale. The court found that they were relied upon here. The misrepresentations were so untrue as to be material and substantial, constituting an adequate ground for rescission. The court noted that the object of rescission is to undo a contract and restore each party to the position it was in before the contract was made. What must be restored? The purchaser's bond and mortgage should be canceled. In addition, the cash paid to Lodge Gate should be repaid to the prospective purchaser, subject to any benefit it may have received from the property while in possession of it. Finally, the prospective purchasers should reconvey the land to Lodge Gate. The court noted that the return of consideration upon rescission is not an award of damages suffered by reason of having been induced to make a contract. It is part of the process of undoing the contract and restoring the status quo.

§ 4.4.3 Excuses

Misrepresentations as to vacancies, rentals, and the status of mortgages resulted in a court ordering rescission in *Nuccio v. 103 East 114th Street Corp.*, 136 Misc. 485, 241 N.Y.S. 333 (Bronx Co. 1930), aff'd 232 A.D. 802, 249 N.Y.S. 909 (1st Dept. 1931). Nuccio, who owned an apartment house in the Bronx, entered into an exchange of that house for certain other houses owned by the realty corporation in Manhattan. Several months elapsed after title closed before Nuccio made a formal offer to restore the property. Rescission may be granted if a timely offer to restore has been made. The complicating factor in *Nuccio* was that one of the buildings purchased was lost in foreclosure.

The testimony at trial was that restoration had been offered and refused. It subsequently became impossible due to the foreclosure proceeding affecting the status of the mortgages on both Manhattan houses.

The court found for Nuccio and granted rescission.

§ 4.4.4 Opportunity to Restore

In an innocent transfer, before the courts will find a breach of an obligation to restore a benefit, the person receiving that benefit must be given an opportunity to make restoration. However, if the transfer is not innocent, and if it is due to fraud or other misrepresentation, the obligation to restore immediately arises.

Therefore, with an innocent transfer, notice of opportunity to restore is necessary. In a fraudulent transfer, notice is implied, and the obligation to restore is immediate.

§ 4.4.5 Effect of Statute

Suppose you are at the wrong end of a fraudulent contract. You want to have the contract rescinded. Does the law require you to make restitution before seeking rescission in court?

The legislatures of various states have grappled with this question. The statutes of many states specify that the person who has received benefits pursuant to a contract based on fraud or other misrepresentations and who subsequently seeks rescission in court need not tender the restoration of his or her benefits prior to a decision by the court.

4.5 DEFENSES

§ 4.5.1 Change of Circumstances

If a right to restitution exists, it is often based upon a wrongdoing: fraud, nonperformance, or breach of contract. Is a change in circumstances sufficient? Only if some basis exists. *Dwelle-Kaiser Co. v. Aetna Casualty & Surety Co.*, 241 N.Y. 464 (1926), involved a dispute between a subcontractor, Dwelle-Kaiser, and the successor surety, Aetna. A brief review of the characters involved is appropriate. Schaeffer Construction Company had contracted with the City of Syracuse to build a public school. A predecessor surety had supplied a performance bond to the City. Dwelle-Kaiser had a subcontract from Schaeffer to furnish the glass and glazing of the school building. In the course of its performance on the school, Schaeffer Construction Company became insolvent. Aetna knew of this condition and procured from Schaeffer an assignment of all monies due or to become due from the City on the contract. Schaeffer continued to work the project. Schaeffer completed the school building after Aetna had taken charge of the company and placed its representatives in charge of the business and as directors of Schaeffer. Schaeffer Construction ordered and directed Dwelle-Kaiser to furnish the glass and glazing for the school according to its subcontract.

Dwelle-Kaiser alleged that at the time of the direction from Schaeffer it was 1) unaware of Schaeffer's insolvency, 2) ignorant that the amount unpaid by the City of Syracuse would be insufficient to pay for the completion of the work, and 3) unaware that Schaeffer's business was being conducted by Aetna. Upon the completion of the contract, Aetna received the money from the City of Syracuse to repay it for monies that Aetna had advanced to complete the contract. Dwelle-Kaiser argued that the acts of Aetna were with the intent and purpose of causing the contractor to build a schoolhouse to be completed without payment to Dwelle-Kaiser for the glass work and to save Aetna harmless from its obligation on the performance bond. Dwelle-Kaiser sought the subcontract price of the glass work. Dwelle-Kaiser complained that Aetna protected its monies advanced to complete the contract by taking an assignment of monies from the City of Syracuse. Instead, Dwelle-Kaiser would have had Aetna warn it that the construction company had become insolvent.

Dwelle-Kaiser was under a subcontract with Schaeffer to furnish the glass for the school. Aetna was under no duty to keep Dwelle-Kaiser informed of the financial condition or of the status of the work and payments. The bonding company was guilty of no wrongdoing or overreaching in protecting itself from loss instead of protecting subcontractors, as long as there was no deception, false representation, or misleading conduct.

In absence of such wrongdoing, the court found for Aetna.

The position of Aetna is not different from that of Schaeffer. What were the relative duties of Schaeffer in the case of insolvency? There was no duty to reveal its insolvency. The fact of the insolvency would not have relieved Dwelle-Kaiser from performing its subcontract. What Aetna did, it had a right to do. What Dwelle-Kaiser did, it was obligated and under contract to do. The court found no grounds for restitution.

§ 4.5.2 Delay in Rescinding

In attempting to do justice the courts will also consider any delay in seeking rescission of a contract. Why is this? Clearly, a delay in seeking rescission would indicate that the benefits of the contract were satisfactory. The longer you wait to rescind a contract, the more difficult it is for you to argue that you have been aggrieved by that contract.

Suppose you have posted a bid bond or given some other security that, if selected, you will enter into a contract? When that contract is rescinded, what happens to your security?

In *Brendese v. Schenectady*, 194 Misc. 150, 85 N.Y.S.2d 856 (Schenectady Co. 1947), aff'd 273 A.D. 831, 75 N.Y.S.2d 884 (3rd Dept. 1948), aff'd 297 N.Y. 965 (1948), Brendese was a bidder on a municipal sanitary sewer construction project. Pursuant to the bid requirements, Brendese had deposited with the City certified checks in the amount of $5,000.00. Brendese's was the lowest bid. He was awarded the contract and notified of such award. He neither executed the contract nor provided the required performance bond.

The notice to bidders, which required the $5,000.00 deposit, stated that the failure to execute the bond and contract within a specified time period would allow Schenectady to retain the deposit as liquidated damages. The aggregate of Brendese's bid was approximately 50 percent of the next lowest aggregate bid. In preparing his bid, Brendese had estimated fair and reasonable value and cost of trench excavation at a price of $3.00 a cubic yard. Through error and mistake, Brendese failed to include or take into consideration the value and cost of dewatering the excavation for trenches, although he did intend to take such omitted items into consideration as part of the bid price. Brendese claimed that his error and omission materially and substantially affected the amount of each of his bids. In view of the error and omission, Brendese argued that he could not furnish the labor and materials incident to the contract without a great and serious loss to himself.

Brendese had been present at the bid opening and, concerned by the difference between his bid and the others, he checked his worksheets. At that point he discovered the omission from the estimate,for dewatering the trenches. This item would have added an additional $3.50 per cubic yard of trench. Brendese unsuccessfully attempted to obtain a performance bond. During this period, he advised Schenectady officials of his delay in obtaining the bond. Schenectady extended his time to file the bond and enter into the contract.

However, from the date he discovered his error, he never advised any official that he had made an error or mistake in his bid. Brendese admitted that within one month from the date the bids were opened, he had determined that he could not perform the contract without loss. Brendese did not ask to withdraw his bids or be relieved from the ground of mistake.

The court found that Brendese's testimony as to the error in his bid was credible. The error was made without any attempt to defraud or overreach the City. Brendese was entitled to rescission. However, the question of whether or not Brendese is to receive its bid deposit is a question of fact. Schenectady maintained that in the circumstances of the case, retention of the deposit as liquidated damages was justified.

Brendese argued that since Schenectady failed to prove any damages from his failure to enter into the contract and from his delay in advising his reasons for not doing so, Schenectady is not entitled to retain his bid deposits. Furthermore, Brendese argued, the discrepancy between his bid and the other bids submitted alone would charge Schenectady with knowledge of his error. The court noted that, where an owner is charged with knowledge of the bidder's error, those errors are not only gross but are evidenced on the face of the bids themselves. The errors may be in the extension and carrying out of totals. Or the basic unit price may have been misstated; i.e., \$1.50 per cubic yard instead of \$15.00 per cubic yard. The court found that Brendese's mistake did not result in a gross discrepancy. Nor was Brendese's mistake clear on the face of the bid. The court would not assume that Schenectady had knowledge of the inadequacy of Brendese's figures merely by reason of the contrast between his bid and other bids. In addition, where bid security is retained, bidders must promptly make known to the owner their discovery of error. Prompt notification leaves the owner in the position to go forward with an award to the next lowest bidder.

At this juncture, the court had determined that Schenectady could retain the bid deposit. The last question was whether the deposit was a penalty or represented liquidated damages. Liquidated damages were discussed previously at § 4.1.9. If the deposit represented liquidated damages, it could be retained in the absence of proof by Schenectady of actual damages. The amount of the deposit was 5 percent of the amount of the bid. It is generally true that parties to a contract may agree on the amount of damages to be paid upon breach. That agreement will not be overturned unless the amount is grossly disproportionate to the actual injury. The court found that the amount of deposit was of the usual and normal percentage. The court noted that aside from the damages to the City in extra costs involved in readvertising bids and reletting a contract, there are also involved in the proceedings for the making of contracts for public improvements the elaborate preparation of details. The public interest mandates prompt construction of sanitation works. Notices to bidders to compete for public construction are prepared in such a way as to protect the City from default on the part of the bidders and the expense incident thereto. The court found that it was competent for these parties to enter into a provision for liquidated damages.

The court noted that a contractor seeking rescission is not permitted to play "fast and loose." Delay or vacillation are fatal to his or her rights. Had Brendese promptly notified Schenectady of his mistake, the City would have been in a position to permit him to withdraw or to reject and to award the contract to another bidder. The court noted that there was no showing by Brendese that such restoration could have been accomplished at the late date when he first notified Schenectady of his mistake by beginning this litigation. As a matter of common knowledge, the court found that at that late date, the City had no alternative but to readvertise the bids on this work with the attendant additional expense and uncertainties. The City is entitled to the bid deposit.

§ 4.5.3 Ratification of Transaction

As we have seen in *Brendese*, a delay in complaining can hinder your chance for recovery. In *Port Chester Electrical Constr. Corp. v. Hastings Terraces, Inc.*, 284 A.D. 966, 134 N.Y.S.2d 656 (2nd Dept. 1954), the court found that a delay in seeking to rescind a contract would bar restitution since the delay could indicate that the transaction had been ratified. Port Chester Electrical was a subcontractor seeking to recover from the owner, Hastings Terraces, on two notes. Hastings Terraces had executed, endorsed, and delivered to Port Chester Electrical the two notes. The notes were thereafter presented for payment and paid by Hastings Terraces. Hastings Terraces argued that the notes were executed, delivered, and paid under duress. Hastings Terraces sought return of the monies paid.

The court noted that a contract procured by duress and exercised by means of threat is not void but voidable. If you seek to repudiate such a contract on the grounds of duress, you must act promptly, otherwise the law will deem you to have elected to affirm the contract. Hastings Terraces did not claim duress until a year and a half after the alleged duress terminated. The court found for Port Chester Electric, holding that the failure of Hastings Terraces to assert a claim of duress for such an unreasonable length of time constituted waiver of the claim and affirmation of the contract.

4.6 FRAUD

§ 4.6.1 Definition

As we have seen, the majority and almost universal rule is that contracts based on fraud may be voided with recovery of the items exchanged and a restoration of the contracting parties to their respective positions before the contract. But precisely what is fraud? In exploring that question, we can consider particularly whether a subcontractor has a duty to disclose material information to an owner. *City of New York v. Kalisch-Jarcho, Inc.*, N.Y.L.J., June 13, 1991, p. 27 (N.Y. Co. 1991) involved the reconstruction of the slate and refrigeration system at an ice rink. In connection with its contract with the City of New York, Kalisch-Jarcho hired Holmsten Ice

Rinks, Inc. as its subcontractor for the purpose of installing a Rinkmaster system, which was Holmsten's patented ice making system. The Rinkmaster system was installed. Subsequently, refrigeration piping began to rust and leak, causing cracks in the rink's slate. The cracking became so bad that the rink was rendered inoperative. The City made unsuccessful attempts to repair the rink. Finally, the City was forced to rebuild the rink at great expense. The City sought compensation from both its contractor, Kalisch-Jarcho, and from Kalisch-Jarcho's subcontractor, Holmsten, alleging that during design and construction phases of the rink, Holmsten became aware of serious problems with dozens of other rinks equipped with the Rinkmaster system. The City argued that this information was fraudulently concealed from the City. In order for fraud to exist, it must be based on the intentional suppression of a material fact. In addition, the person alleging fraud must show that a duty to disclose had existed.

How does such a duty arise? The court reviewed a number of ways. The parties may be involved in a fiduciary relationship, or one party may possess superior knowledge not readily available to the other, with the other party acting on the basis of mistaken knowledge. A fiduciary relationship may be invoked where there is a contract between the parties. The court found that the lengthy negotiations between Holmsten and the City were purely commercial and at arm's length. There was no relationship of trust or confidence. The City, as a sophisticated business entity, could have discovered for itself any of the alleged defects in other rinks equipped with the Rinkmaster System. The court found for Holmsten and dismissed the City's fraud claim.

CITATIONS

Abner M. Harper, Inc. v. Newburgh, 159 A.D. 695, 145 N.Y.S. 59 (2nd Dept. 1913) 4.4.1

Arnot v. Pittston & Elmira Coal Co., 68 N.Y. 558 (1877) 4.2.7

Brendese v. Schenectady, 194 Misc. 150, 85 N.Y.S.2d 856 (Schenectady Co. 1947), aff'd 273 A.D. 831, 75 N.Y.S.2d 884 (3rd Dept. 1948), aff'd 297 N.Y. 965 (1948) 4.5.2, 4.5.3

Callanan v. Powers, 199 N.Y. 286 (1910) 4.1.4

City of Elmira v. Larry Walter, Inc., 76 N.Y.2d 912, 563 N.Y.S.2d 45 (1990) 4.1.9

City of New York v. Kalisch-Jarcho, Inc., N.Y.L.J., June 13, 1991, p. 27 (N.Y. Co. 1991) 4.6.1

Cranford Co. v. New York, 150 A.D. 195, 134 N.Y.S. 839 (1st Dept. 1912), aff'd 211 N.Y. 534 (1914) 4.1.6

CHAPTER 5

MISTAKE, DURESS, THREATS, AND UNDUE INFLUENCE

5.1 MISTAKE

§ 5.1.1 Mutual or Unilateral Mistake

Two definitions are in order. Mutual mistake is a situation where both parties to the contract are mistaken as to some essential element. In that circumstance the courts will allow rescission. A unilateral mistake involves a mistake by only one of the parties to the contract. Rescission in that situation is only allowed where there has been some wrongdoing, fraud, misrepresentation, or deceit by the other party.

We have discussed a contractor's need to maintain comprehensive builder's liability insurance. While a building is under construction, you are responsible for that building. If it should be lost to fire, it is your loss. But suppose the building is about to be transfered to another by its owner; if a fire occurs, who bears that loss? A similar situation is discussed in *Powderly v. Aetna Casualty & Surety Co.*, 72 Misc.2d 251, 338 N.Y.S.2d 555 (Monroe Co. 1972). The Arcieris were to purchase property from the Powderlys. Closing was set for January 15. At 10:00 A.M., Powderly and his attorney attended the closing. The Arcieris' attorney was also present, although they were not. The deed, executed by the Powderlys was given to the Arcieris' attorney. The Arcieris' attorney then gave the Powderlys' attorney a check with the direction to hold it until the title abstract was redated, showing clear title and papers were recorded. The Arcieris' attorney would notify the Powderlys' attorney once this was accomplished. The Arcieris executed the deed at approximately 1:00 P.M. Their attorney attempted to contact an insurance agent, but was unsuccessful until 3:15 P.M., when he arranged for an immediate insurance binder for the Arcieris. The first report of a fire at the premises was turned in via a fire alarm box at 3:19 P.M. These are facts that delight litigators and dismay bystanders.

There are three major questions here, only one of which concerns us for this discussion. The questions are who was the owner of the property? If Arcieri was the owner, was his insurance policy in effect? If Powderly was the owner, was his insurance policy in effect?

The court first dealt with the question pertinent here. Was there a mutual mistake of fact between the Arcieris and their insurance carrier. The binder was given by the insurance company to Arcieris' attorney at 3:15 P.M. The box alarm was rung in at 3:19 P.M. The court found that the onset of the fire had to precede the alarm by more than four minutes. The court noted that the binder between the Powderlys and the Arcieris contained a reservation that if a fire occurred before closing, either party had the option of being relieved. If at the time the binder was given a fire had started, Arcieri could have availed himself of the escape clause because damage was substantial.

There was a mutual mistake of fact between the Arcieris' and their insurers. Both assumed an insurable building at the time of the insurance binder. Both assumed an insurable risk, which, in fact, turned out to be in the throes of a fire at the time the insurance binder was given. Because of the mutual mistake, Arcieris' insurer is not liable.

Not to leave you hanging, the court then found that there had not been delivery of the deed due to conditions placed upon its transfer to the Arcieris' attorney. Therefore, at the time of the fire, the Powderlys were owners of the property. Since the Powderlys did have insurance on the property, Aetna was responsible and must pay for the fire loss.

§ 5.1.2 Mistake as to Identity

You have a right to determine the person with whom you contract. If you are mistaken as to the identity of that person, there can be no meeting of the minds and there can be no contract.

Kelly Asphalt Block Co. v. Barber Asphalt Paving Co., 211 N.Y. 68 (1914), involved a dispute between Kelly and Barber, who sold certain paving blocks to Kelly's agent. Kelly was suing as an undisclosed principal, and Barber contested its right to do so. The general rule is that a contract made in the name of an agent, as ostensible principal, may be sued on by the real principal. Barber argued that the identity of the principal had been concealed because of the belief that were it disclosed, the contract would not be made. Kelly and Barber were competitors. Kelly's president suspected that Barber might refuse to name him a price for the paving blocks. Suspicion was not based on a previous refusal, but solely on their relationship as competitors. Kelly engaged the services of Booth. Booth asked Barber for a price, received a quotation, and the asphalt paving blocks were ordered in Booth's name. The blocks were delivered, and payment was made by Booth with money furnished by Kelly. The paving blocks were unmerchantable. Barber retained the price and argued that if it had known that Kelly was the principal, it would have refused to make the sale.

The court noted that a contract involves the meeting of the minds of the contracting parties. If one of the supposed parties is wanting, there is an absence of such meeting. In such a case, there is no contract. Cases have been reported where A has ordered the merchandise of B, and C has surreptitiously filled that order. Can C, having thrust himself, without consent, into the position of a creditor be entitled to recover the value of his wares? The court implied no, but did not specifically address that issue. The court stated this proposition to accentuate the distinction between the cases that involve it and the case at hand. In the case at hand, neither of the supposed parties were wanting. The meeting of the minds between the contracting parties, Booth and Barber, was real not illusory. Barber was contracting with the precise person with whom it intended to contract. Barber gained whatever benefit it may have contemplated. An agent contracting in his or her own name for an undisclosed principal does not cease to be a party because of his or her agency. Such an agent, having made himself or herself personally liable, may enforce a contract although the principal has renounced it. There is a distinction where the contracting party is an agent for a disclosed principal. In such a case, the agent himself or herself is not personally liable. Booth, as between himself and Barber, is liable as principal to the same extent as if he had not been acting for Kelly.

The court found that it was impossible, under these circumstances, to hold that the contract collapsed for want of parties to sustain it. Here, two persons were bound, and those were the very persons intended. Had Booth given the order in his own right and for his own benefit but expecting later to assign it to Kelly, that undisclosed expectation would not have nullified the contract. Similarly, his undisclosed intention to act for a principal who was unknown to Barber was equally ineffective to destroy the contract. The unsuspected existence of an undisclosed principal can supply no grounds for the avoidance of this contract unless fraud is proved.

The court then distinguished between a mistake as to identity, which renders a contract void from the beginning, because contractual ties have never been completely formed; and a contract existing on the basis of a fraud, which renders the contract voidable at the election of the defrauded party. If a person who is in reality an agent denies that agency and falsely asserts that his or her principal has no interest in the transaction, the contract becomes voidable, not because of a want of parties, but because it has been fraudulently procured. The court found no misrepresentation in the case at hand. Booth had made no misrepresentation to Barber. He was not asked, nor did he say anything concerning Kelly's interest in the transaction. Barber would have had the court determine that the validity of a contract turns not on any overt act of either Kelly or Booth, but on the presence or absence of a mental state. Barber had asked the court to hold that a contract complete in form becomes a nullity in fact because of a secret belief in the mind of the undisclosed principal and that disclosure in his name would be prejudicial to completion of the bargain. The court refused to go so far. Barber was entitled to damages it incurred because of the unmerchantibility of the asphalt paving blocks.

It is curious that the decision in *Kelly* was written by Judge Benjamin N. Cardozo. His succinct and precise style has endeared Judge Cardozo to generations of attorneys, but some 14 years before, Judge Cardozo was attorney Cardozo, representing John Weber & Co. In *John Weber & Co. v. Hearn*, 49 A.D. 213, 63 N.Y.S. 41 (1st Dept. 1900), there was no question that a contract was made with Weber. The question, though, was whether the contract was made with a partnership or a corporation. The proof had clearly shown that Weber was a corporation and not a partnership. Hearn had not claimed that any representations were made during negotiations to the contract that Weber was a partnership. The work had been actually performed by the corporation. In the absence of anything showing that Hearn had been misled or injured by the fact that Weber was a corporation and not a partnership, Hearn had no legal grievance. The court found that the fact that Hearn thought he was dealing with a partnership would not be a bar to Weber's right to recover for the value of the work on this building contract. The fact that Hearn thought he was dealing with a partnership did not prevent Weber's right to recover the value of the work done by it as a corporation.

In that more gentle age, it is not surprising that neither of the attorneys in *Kelly* cited this case to Judge Cardozo's court. I cannot help but feel that 14 years later Judge Cardozo was reviewing in his own mind the very arguments he had mustered on behalf of Weber when he made his decision in *Kelly.*

§ 5.1.3 Legal effect

If the language of a contract is clear, the fact that a contracting party misinterprets the legal effect of that language will not entitle that party to overturn the contract. The contracting party cannot avoid that contract by arguing that the terms were misunderstood. Rescission will not be allowed as to a party's erroneous opinion as to the legal effect of the document. Where material facts have been disclosed and an opportunity to make an independent determination as to the legal effect of those facts, a person cannot be held allowed to complain of cognizable misrepresentation. This remains the case even if that person has expressed an erroneous opinion as to the legal effect of the facts. *Jacob Goodman & Co. v. Pratt*, 138 N.Y.S.2d 89 (N.Y. Co. 1954).

§ 5.1.4 Ratification of Mistake

When you discover a mistake you must immediately notify the owner. If notification is not immediate, the court may find a waiver of the mistake or a ratification of it. What is the rule where the owner has made a mistake and does not reveal it to you, the contractor?

Wholesale Service Supply Corp. v. State, 201 Misc. 56, 103 N.Y.S.2d 820 (Ct. Cl. 1951), involved a contract for the purchase of a specified amount of fencing posts and accessories. It was the owner who did not reveal his mistake. Upon award of the contract, Wholesale Service placed an order for the fencing, which was not a stock item, but was specially fabricated. Wholesale requested that the State issue a delivery order for the fencing. Wholesale was directed to a deputy commissioner who stated that it required only 60 rolls of fencing rather than the 600 rolls stated in the proposal and award. The State had received authorization and subsequently placed delivery orders from 120 rolls of fencing. The State refused to accept any additional rolls. Wholesale managed to dispose of some of the rolls of fencing. It was unable to dispose of 437 rolls of fencing, which the manufacturer refused to take back.

The court had reviewed the proposal, the award, and the contract. It was apparent at the time of bid and at the time the award was made that each party believed that it was contracting for a specified quantity. An error had been made by one party to the contract, the State, but Wholesale was completely unaware of that error. The error was made through the negligence of an agent of the State. The court noted that in order to relieve the State from the effects of a unilateral mistake, it is imperative that the innocent party who had no knowledge of the mistake, Wholesale, and no reason to suspect its existence, must not have been prejudiced by any change in position made in reliance on the contract. The court found that Wholesale had no knowledge of the error. There was no contract requirement for Wholesale to check the requisition on file to assure itself that no clerical error had been made in designating the quantity in the proposal and award. There was nothing on the face of the contract to put Wholesale on notice that an error had been made, nor was there any uncertainty that would have required Wholesale to satisfy itself that the State had not been negligent in preparing the proposal and award. Parties to a contract are presumed to deal in good faith with each other. Wholesale had a right to rely on the fact that the State, in preparing the proposal and award, would do so carefully and prudently.

The court found that Wholesale did not learn of the mistake until late December. Prior to that time, all fencing material ordered from the manufacturer had been delivered. The record shows, however, that some time between October and December, the State became aware of the error and did nothing to apprise Wholesale. Such conduct was not commensurate with the good faith required.

The court found an unreasonable delay on the part of the State. Furthermore, the failure of the State to notify Wholesale after discovery of the mistake constituted a ratification of the mistake. Where a party desires to rescind a contract on the grounds of mistake or fraud, he or she must upon discovery of the facts at once alert the other party to the contract. Silence will waive any objection as surely as if the mistake or fraud had not occurred. Delay and vacillation are fatal to the right to rescind. The court found for Wholesale.

5.2 DURESS

§ 5.2.1 By Third Person

As a general rule, a contract is not affected by duress exercised by a stranger to the contract. What does a court consider duress? Certainly a gun to the head will do it. In *Austin Instrument, Inc. v. Loral Corp.*, 29 N.Y.2d 124, 324 N.Y.S.2d 22 (1971), the concept of economic duress was explored by the court. Where a contract is executed due to duress from business necessity or compulsion, the party coerced is excused from performance. Loral was awarded a $6 million contract by the Navy for the production of radar sets. A typical government contract, it contained a schedule and liquidated damages for late deliveries. In the event of a default by Loral, the contract contained a cancellation clause.

Loral solicited bids for some 40 precision gear components needed in the production of the radar sets. Austin was awarded a subcontract to supply 23 such components. Loral was awarded a second government contract and again solicited bids for 40 precision gear components. Loral informed Austin that it would be awarded a subcontract for those items on which it was the low bidder. Austin refused to accept an order for less than all forty gear components. Austin then informed Loral that it would cease delivery for all parts under the existing subcontract unless Loral consented to substantial increases in price, retroactively for parts delivered and prospectively for parts not yet shipped; and also subcontracted with Austin for all 40 parts needed under the second Navy contract. Austin subsequently stopped delivery. The court found that Austin's threat to stop delivery unless the prices were increased deprived Loral of its free will.

Loral contacted ten manufacturers of precision gear components and found none who could produce parts in time to meet Loral's commitment to the Navy. Loral then acceded to Austin's demands in a letter that stated, "...we are left with no choice or alternative but to meet your conditions." Loral agreed to the price increases insisted upon by Austin in the first subcontract and awarded Austin the second subcontract for all 40 precision gear components.

At issue was neither the facts nor the law, but the application of the law to the uncontested facts. The applicable law is clear. A contract is voidable on the grounds of duress when the organization claiming duress establishes that it was forced to agree to a contract by a wrongful threat that precludes the exercise of its free will. Economic duress or business compulsion is established by proof that the immediate possession of needful goods is threatened. The mere threat to breach a contract by not delivering required items, however wrongful, is not in itself economic duress. The threat must also be joined with the inability of the threatened party to obtain the goods elsewhere and the inadequacy of the ordinary remedy by action for breach of contract.

Because of the production schedule in its contract, Loral had to plan ahead. Substantial liquidated damages plus the threat of default were genuine possibilities. A substantial portion of Loral's business was with the government. Loral had a real concern that failure to deliver on these contracts would jeopardize its chances for future contracts. These concerns are not self-imposed, undisclosed, and subjective. The court found it reasonable for Loral to consider itself in an emergency situation.

Austin, on its part, argued that Loral should have contacted the government and asked for an extension of time to enable it to purchase parts from another vendor. The court disregarded this argument. The only promise that Loral received from the companies it contacted was for commencement of delivery, not full supply. It would be impossible to know the length of the extension of time to be requested. In addition,

Loral was bound by controlling authority that nonperformance by a subcontractor is no excuse for default in a government contract.

Loral demonstrated that it could not obtain the parts elsewhere in a reasonable time. Loral was producing a highly sophisticated item of military machinery. The component parts needed to be of the strictest engineering standards. Loral did not need to go to vendors, other than the ten it had contacted, with whom it was either unfamiliar or dissatisfied. Loral had contacted all manufacturers it believed capable of making the components.

Austin argued that Loral, even if it did enter into the subcontract under duress, lost its rights for a refund by waiting long after the termination date to disaffirm it. It is, of course, true that to recover on the ground of duress, that claim must be made promptly. Yet here, Loral delayed in making its demand for a refund until three days after Austin's last delivery on the second subcontract. Loral's reason was that it feared another stoppage of delivery. Considering Austin's conduct in the past, the court found this perfectly reasonable. The court found for Loral.

Note that duress exerted by a third party to a contract will not allow for a contracting party to avoid the contract. Suppose that third party is acting with the knowledge of one of the contracting parties. In that situation, the contract may be voided at the option of the contracting party claiming duress.

§ 5.2.2 Against Third Person

The general rule is that a contract cannot be avoided because of duress exerted on a third person. The law will not consider a person under duress who enters into a contract to relieve another person.

The principle is illustrated in *George Colon & Co. v. East 189th St. Bldg. & Constr. Co.*, 141 A.D. 441, 126 N.Y.S. 226 (1st Dept. 1910). A surety company that had posted a bond to discharge a mechanic's lien argued that the contract between Colon and the owner of the property, East 189th, was tainted by fraud and duress. Colon had threatened to abandon the work on the premises and to induce others to abandon their work unless the owner would make the contract now sued upon. The court held that the fraud and duress defense is unavailable to the surety company. A contract procured by duress and by means of threats is not void but voidable. The general rule is that only the party upon whom the duress was exercised may take advantage of it to avoid the contract. The right of avoidance is personal to the party coerced.

The court noted that a surety on a contract that was coerced by duress, may plead that defense if it was ignorant of the duress when it became surety. However, this exception does not apply to the surety company here. The surety company was not a surety on the voidable contract, as for example, had the surety company posted performance and payment bonds, but was rather the surety on the bond to discharge the lien. As such, the surety cannot have any defense that would put Colon, the contractor, in a worse position than it would have been in if the lien had remained undischarged by the bond. If no undertaking had been given to discharge the lien, the right to avoid the contract because of duress would still rest solely with the owner.

The court went further and noted that even if the surety company's argument was correct, it cannot be permitted to avoid the contract. One who would repudiate a contract on the grounds of fraud or duress must act promptly or will be deemed to have elected to affirm it. Fifteen months had elapsed from the making of the contract. Even the owner, East 189th, would not have been permitted at that date to repudiate the contract. Finally, the court noted that the surety company's own contract, upon which it was now sought to be held liable, had not been alleged to be tainted with fraud.

There are, of course, exceptions to this rule. If a husband or wife is induced to execute a contract, because of duress on a spouse, such a contract may be avoidable. The courts hold the same to be true where parents are induced to execute contracts because of duress on their children.

§ 5.2.3 Economic Compulsion

Economic duress was touched upon at § 5.2.1. *Stewart M. Muller Constr. Co. v. N.Y. Tel. Co.*, 40 N.Y.2d 955, 390 N.Y.S.2d 817 (1976), involved an attempt to declare a settlement agreement null and void. The court noted that any contract may be voided on the grounds of economic duress, where the complaining party was compelled to agree to the terms by means of a wrongful threat that precluded the exercise of free will. Muller alleged that the settlement agreement was induced by New York Telephone's threat to terminate an earlier contract. Muller, however, did not allege that New York Telephone was not within its contractual rights in threatening to exercise the termination clause contained in that contract. The court noted, rather, that New York Telephone, in the context of a contractual dispute, preserved its right by following the letter of the termination clause. At the same time, New York Telephone sought an accommodation with the financially hard pressed Muller. In view of the explicit provisions of the termination clause giving New York Telephone the right to cancel the contract upon an architect's certification of substantial breach, there is no possibility that Muller could present evidence that would establish that New York Telephone's threatened cancellation was in excess of its contractual rights. Such an exercise cannot be deemed wrongful, therefore the court found for New York Telephone.

§ 5.2.4 Effect of Ratification

Even if an agreement is entered into under duress, the party subject to the duress may subsequently ratify that agreement. *J.R. Constr. Corp. v. Berkeley Apartments, Inc.*, 259 A.D. 830, 19 N.Y.S.2d 500 (2nd Dept. 1940). A person who wishes to repudiate a contract executed under duress must act promptly or will be deemed to have elected to ratify that contract.

5.3 THREATS

§ 5.3.1 Element of Coercion

One of the essential elements of a threat is coercion. Before a court will consider an action to be a threat, coercion must be proved. A court will look for proof of improper pressure or influence that destroyed the free agency of a party and caused that party to make a contract not of its own volition. In *Scolaro v. Ensign Improv. Co.*, 169 N.Y.S. 472 (App. T. 1918), Scolaro sought to recover on a written contract for the performance of marble work and on a second contract for extras. Ensign, a developer, had been engaged in the erection of a five-story apartment building. Ensign had purchased marble in Massachusetts for the bases, treads, and risers of the steps. Work on the building had been performed on a Saturday and Sunday in violation of union rules. The marble work had not been completed, but everything had been done with the exception of installation of the ground floor marble work. This work consisted of the construction and setting up of 50 feet of bases and the treads and risers for the first flight of stairs. Eighty-five percent of the work had been completed when the mechanics working on other parts of the building were prohibited by their union delegates from proceeding with their work.

Ensign's President met with Scolaro. Scolaro told him Ensign could not proceed with the building unless Scolaro was given the contract for marble work. Scolaro stated that Ensign would not have to give Scolaro the job, but if the job was to go ahead, he would have to contract with Scolaro.

Ensign's president signed a contract in his office a few days later, and the marble work went forward. Subsequently, Ensign's president objected to the marble work being all taken down and reset under the contract. It was agreed that Ensign should pay Scolaro to reimburse him for loss in payment of wages for men hired by Scolaro in anticipation of Scolaro doing the job. Ensign paid the amount to Scolaro.

Scolaro then stated he was short some marble. When Ensign's president protested, Scolaro said if he wanted the marble, he would need to execute a second contract for extra work in order to get the job finished.

When Scolaro sought reimbursement on the contracts, Ensign argued duress.

The court found that the initial contract for marble work with Scolaro was signed by Ensign's president in his own office. The signing occurred a few days after the conversation with Scolaro. The court found this situation precluded Ensign from claiming that its president's act was not an act of his own volition and that he was incapable of exercising free will. With respect to the extra payment for lost wages and for extra marble, the court found substantially the same. Both contracts were performed by Scolaro, therefore, the court found for Scolaro and ordered payment by Ensign.

Realize that in 1918,the concept of economic duress was not as highly developed as it certainly is now. The court in *Scolaro* appears to be looking for a gun to the head type threat. Certainly, a court considering the realities of labor unions in the construction industry could find that Ensign was cut off from his supply of a very important component in any construction project, his labor.

§ 5.3.2 Threat to Break Contract

We have seen that the threat to break a contract, while wrongful, does not alone meet the legal threshold for duress, allowing for a voiding of the contract. *30 East End Inc. v. World Steel Products Corp.*, 110 N.Y.S.2d 754 (Bronx Co. 1952), involved the construction of a 108-family apartment house and the materials to be furnished in accordance with the terms of the contract. World Steel was to supply interior steel door bucks, steel window linings, and other appurtenances not readily available or procurable on the open market. It was alleged that World Steel falsely represented that it would not deliver the materials and would not allow the furnishing by Pioneer Fireproof Door Corp. of materials unless a modification and rescission agreement, which relieved World Steel of the obligation to deliver said materials, was executed. The owner complained of this pressure, but having no immediate means of relief, it agreed to the modification and paid involuntarily to Pioneer Fireproof Door, a higher price for the material, sustaining the damage sought. The owner alleged duress by World Steel and also charged that World Steel and Pioneer Fireproof Door entered into a scheme to defraud and deceive it. The crux of the owner's claim was that World Steel's refusal to deliver and to consent to have Pioneer deliver, taken together with the scarcity of material in the open market, resulted in economic duress. The court noted that there were no facts to indicate that the threat not to allow Pioneer to deliver was coercive. The owner had not alleged that World Steel possessed or was believed by the owner to possess power to prevent Pioneer from delivering the material. A lack of such power would render the threat to withhold consent innocuous. In the absence of such a belief by the owner, it cannot be said that he acted involuntarily. Nor was it alleged that the owner could not have bought the material directly from Pioneer without World Steel's consent. In addition, there was no statement that the material was not readily procurable nor that the owner had made any efforts to buy it on the open market.

This appears to be the fact that distinguishes the result in this case from the result in *Austin Instruments*, § 5.2.1. There must be no other means of relief. The coercion and duress must be sufficient to result in involuntary compliance. The court noted that while the owner may have needed the material for the orderly construction of its building, World Steel, to protect its own interests, had the right to drive a hard bargain, provided its acts were lawful. Whatever the fear of economic loss, it must be induced by some unlawful act. The court found for World Steel.

§ 5.3.3 Communication of Threats

It is not necessary that the notification of threatened duress come from the actual person who will benefit. The communication of a threat through a third party is sufficient if the person benefiting knowingly avails himself or herself of such threat.

§ 5.3.4 Lawfulness of Threatened Act

Remember, where a threatened act is perfectly lawful, there can be no duress. In Muller Constr., § 5.2.3, actions within and allowed by the terms of the contract could not form a basis for duress or threat.

E.M. Substructures, Inc. v. City of New York (Sewers 149th St., Queens), 73 A.D.2d 608, 422 N.Y.S.2d 444 (2nd Dept. 1979), appeal dismissed 49 N.Y.2d 878, 427 N.Y.S.2d 990 (1980), involved a construction contract with the City of New York. E.M. sought recovery for extra work. In exchange for an extension of time to complete the contract, E.M. had agreed to waive all claims. E.M. did not dispute that it could have reserved its claims by excepting them from the waiver. E.M. conceded it was aware of the claim for extra work when it signed the waiver. On past occasions, E.M. had preserved such claims by excepting them from such waivers by written side agreements. The court found that by signing the waiver without exceptions, E.M. forfeited its claim. Having obtained the benefit it sought, the extension of time, E.M. cannot now avoid that which it gave in return.

E.M. argued duress contending that the City's conduct precluded it from asserting the waiver. E.M. had presented no evidence of the kind to preclude the City from asserting waiver. Merely requiring a contractor to waive its claims for damages in return for an extension of time does not, without more, establish duress.

5.4 UNDUE INFLUENCE

§ 5.4.1 Definition

There is no bright line between undue influence and duress. Clearly, undue influence is something less than duress, however. In doing justice, a court, where the facts are insufficient to constitute duress will rely on the ground of undue influence. The court's justification is simple. The contract may be avoided since the overpersuasion present in undue influence amounts in law to coercion. The courts will not enforce a coercive contract.

CITATIONS

CHAPTER 6

MECHANIC'S LIENS

6.1 GENERALLY

The most obvious question is, why is a discussion of mechanic's liens contained in a book discussing contract clauses and case law? I could easily point out that many contracts contain provisions for the contractor, on payment, to sign a "waiver of lien" document and leave it at that. But that is not all. The fact is that mechanic's liens must be discussed. The mechanic's lien is very often the most expeditious and least costly means available to a contractor or subcontractor to obtain payment.

The principal difficulty with discussing mechanic's liens on a national basis is that mechanic's liens are creatures of statute. This means that in each of the 50 states, the legislature establishes the framework within which a mechanic's lien can arise. Our discussion here is limited to three states: New Jersey, New York, and Texas. A caution is in order. This is, then, a "bare bones" discussion. It is not meant to be comprehensive, nor should it be construed as even being current. As I write, some legislature somewhere may be changing the mechanic's lien law as stated here. I will touch upon the complexities and procedural pitfalls involved. But my advice to you, in *this* more than in any other area of contract law, is to seek the counsel of a local attorney.

Since a lienor's rights are statutory, they are construed strictly. A seemingly insignificant deviation from the statutory requirements can be fatal.

Consider this New York example. In New York, the lien rights of contractors and subcontractors are contained in the Lien Law enacted by the New York State legislature. For years, the lienor needed only to file various information required by the Lien Law. The filings would be made, in the case of a private lien, with the County Clerk in the county in which the property was located. Some clarification: a private lien attaches to the property itself. A public lien, on the other hand, arises where a government or governmental agency owns the property. Since it is owned by the government, the lien cannot attach to the property. Rather, the lien attaches to a fund of public monies available for the construction. In New York, a public improvement lien is filed with the head of the particular governmental agency performing the construction and with his Chief Financial Officer. This was a straightforward procedure. Many contractors and subcontractors took it upon themselves to make the necessary filing.

I've warned you of legislative changes affecting mechanic's liens. In 1988, the legislature put in effect specific service requirements in connection with both private and public liens. The failure to comply strictly with those requirements would terminate a lien. For private liens, service of a copy of the lien was required on both the owner and the entity with whom the lienor had his or her contract or subcontract. A lienor having a contractual relationship with either a subcontractor or a subsubcontractor, but not with a contractor who had a direct contract with the owner, must also serve the lien, by certified mail, on the contractor.

And how is service on the owner effected? Within 30 days after filing the lien, the lienor must serve the owner, if not a corporation, by delivering the lien to him or her personally; if he or she cannot be found, then by delivering the lien to his or her agent or attorney, or by leaving at the owner's last known residence in the city or town in which the property liened is located with a person of suitable age and discretion. If the owner cannot be found, the lien can be sent, by registered or certified mail, to the owner's last known place of residence. If the owner has no such residence in the city or town in which the property lien is located, or cannot be found, or has no agent or attorney, a copy of the lien may be affixed to the property between the hours of 9:00 A.M. and 4:00 P.M.

Now, if the owner is a corporation, service of the lien must be made by delivering a copy of the lien to the president, vice president, secretary or clerk of the corporation, the cashier, treasurer, director or managing agent of the corporation, personally within the state, or, if none of those officers can be found within the state, by affixing a copy thereof to the property between the hours of 9:00 A.M. and 4:00 P.M., or by registered or certified mail, addressed to the corporation at its last known place of business. If the service requirements are not complex and difficult enough, the law also requires that, within 35 days after filing the lien, the lienor must file proof of service of the lien with the County Clerk.

In addition to serving the owner, within 30 days after filing the lien, the lienor must serve a copy of the lien, by certified mail, on the contractor or subcontractor by whom he or she was employed or to whom he or she supplied materials. If the lienor has a direct contractual relationship with a subcontractor or a subsubcontractor, but not with the contractor of the owner, he or she must also serve a copy of the lien, by certified mail, on the contractor. Again, proof of service upon the person by whom the lienor was employed or to whom he furnished materials, and upon the contractor, must be filed with the County Clerk within 35 days after the lien was filed.

The 1988 amendments also affected public liens. Simultaneously with filing the public lien, the lienor must serve a copy, by certified mail, on the contractor or subcontractor by whom the lienor was employed or to whom he or she furnished materials. The lienor having a contractual relationship with a subcontractor or a subsubcontractor, but not with the contractor, shall serve a copy of the lien, by certified mail, on the contractor. Proof of service is filed with the lien.

In 1989, the first case arose concerning these amendments. *Paolangeli v. Sopp,* 145 Misc.2d 259, 546 N.Y.S.2d 322 (Tompkins Co. 1989), involved a private lien. The lienor filed the lien with the County Clerk and served the owner the following day. However, proof of service was never filed within 35 days. The court found that the failure to file the proof of service was not a mere irregularity, which could be corrected. The court held that the 1988 amendment was clear and unambiguous. No aid was needed from the court for its interpretation. The plain and clear effect of the statute is that the mechanic's lien terminates 35 days after filing of the lien unless proof of service has been filed. The lienor's failure to file the proof of service within 35 days cannot be excused. The lien had, therefore, terminated.

§ 6.1.1 Background

As I cautioned above, this will be a "bare bones" discussion. It is not meant to reflect the present state of the law, since a legislature may change the statutory formulation at any time. However, it is as up to date as is reasonably possible. Its purpose is to illustrate the various approaches taken by three states in approaching the issue of mechanic's liens. You will see similarities. You will see vast dissimilarities. The point is, your local attorney will know the current state of the lien law in his or her locale. He or she can guide you and help you to avoid those pitfalls that will destroy your lien rights.

In general, a contractor, subcontractor or laborer can file a lien. In New York, the filing is with the Clerk of the County in which the liened premises are located. A New York lien must be verified, that is, sworn to before a notary. New Jersey has an interesting twist. In New Jersey, if the contract and specifications, signed by the owner, are filed with the County Clerk before work is performed or materials furnished, only the general contractor can lien for those items. All of the laborers and materialmen must proceed by stop notice. Now, remember, New York requires that the owner, contractor, subcontractor, and person with whom the lienor had a contract be served within 30 days, and proof of service filed within 35 days, with the County Clerk. In Texas, the lienor must file an affidavit, again a statement sworn before a notary, claiming the lien, with the County Clerk of the county where the property is located, no later than the 15th day of the fourth calendar month after accrual of the debt. If the lienor wishes a lien on funds retained, he must file the affidavit no later than 30 days after work is complete. In addition, in Texas, after filing the affidavit, the lienor must send a copy, by registered or certified mail, to the owner and original contractor no later than ten business days following filing of the affidavit or a filing deadline, whichever is earlier. If the lienor is other than the original contractor, he or she must also given written notice of the unpaid balance to the owner no later than the 15th day of the third month following each month in which labor is performed or materials furnished. The notice must also be given to the original contractor within the same period. There is, of course, a variation where the lien arises from a debt incurred by a subcontractor.

In New Jersey, a person other than the contractor who files a notice of intention prior to the filing of the contract can lien ahead of the contractor until receipt of written notice, personally or by registered mail, of the time and place of filing of the contract. It must be recognized that, in New Jersey, there can be no lien prior to the filing with the County Clerk of a signed notice of intention to perform labor or furnish materials, containing a description of the land, the name of the record owner, and other information. The notice of intention expires after two years.

When must the private lien be filed? In New Jersey, the lien must be filed with the County Clerk within four months of the last labor performed or materials furnished. In New York, the private lien must be filed within eight months. However, it must be filed within four in the case of a single-family dwelling. There are certain restrictions with respect to filing liens on condominiums. In Texas, architects, engineers, and land surveyors who prepare plans or plots in connection with construction or repair work have a right to lien. However, they must have a written contract with an appropriate property description recorded by the County Clerk of the county where the property is located before construction commences.

In Texas, the owner is required to retain, during the work and for 30 days thereafter, ten percent of the contract price as a fund for lienors. Where the owner has withheld this amount from the original contractor and a lienor has given written notice that a claim is due, the lienor can make a written demand on the owner for payment by registered or certified mail with a copy to the contractor. If the contractor does not inform the owner within 30 days that he or she intends to dispute the claim, he or she is deemed to have agreed to it. The claim must then be paid by the owner to the lienor.

The New York Lien Law contains a trust fund article. Pursuant to the trust fund provisions, construction monies are held in trust for potential lienors. In New Jersey, the owner is protected to the extent of payments made to any lienor in good faith after being served with a copy of a filed notice of intention and is entitled to credit such payments against the general contractor.

In New York, a waiver of lien is void and unenforceable unless the lienor has actually been paid for the labor performed or materials furnished. There should be no concern in your mind if your contract requires that you sign a waiver of lien as a condition of payment. Should you sign that waiver, and the payment never occurs, or the payment check is bad, the waiver is void.

In Texas, a lienor for specially fabricated material must give written notice to the owner and to the original contractor, where his order is from a subcontractor, no later than the 15th day of the second month following the month in which the lienor receives and accepts the order. In New York, no prior notice is necessary, and the specially fabricated materials, even if not delivered to the site, may be included in the mechanic's lien.

A public lien in New York has a life of one year. It may be extended by affidavit for a subsequent year and then by court order for subsequent years. A lien on a single-family dwelling may only be extended by court order.

In New Jersey, a mortgage will have priority over a mechanic's lien where certain conditions are fulfilled. In New York, contracts for building loans must be in writing and acknowledged and, on, or before the date of recording of the building loan mortgage, filed with the office of the County Clerk where the property is situated. Subsequent modifications of the building loan contract must be filed within ten days after execution of such modification. If these conditions are not fulfilled, subsequently filed mechanic's liens will have priority.

In New York, laborers for daily or weekly wages have priority over contractors and subcontractors. In New Jersey, journeymen and laborers have priority for wages over employers, contractors, and materialmen.

There is no priority among liens as of the time of filing in New York. The same is true in Texas.

With respect to public liens, in New Jersey, a verified notice must be filed with a public agency before or within 60 days after, completion and acceptance of the project. In New York, a public improvement lien must be filed with the agency within 30 days of completion and acceptance. Along with the lien, proof of service on a contractor, subcontractor, or person with whom the lienor contracted must be filed. The duration of a public improvement lien in New York is six months. In Texas, the person furnishing materials or labor to a contractor under a public contract not exceeding $25,000 can obtain a lien on the money due to the contractor. If the public contract exceeds $25,000, that person obtains a right against the payment bond of the prime contractor.Finally, in Texas, a court may award costs and attorneys' fees in connection with the foreclosure of a lien.

No one claimed this would be simple. The fact is, your attorney is your best guide in the mechanic's lien maze.

CITATIONS

CHAPTER 7

THIRD-PARTY BENEFICIARIES

7.1 GENERALLY

7.1.1 § Contract Enforcement

When you sign a contract with an owner or a subcontractor, you recognize that you are making yourself liable to that person. But are there other persons to whom you would be liable? Can persons other than the contracting parties enforce a contract?

The clearest way to avoid a liability to third persons is to include a contractual provision that specifically declares that the contract in no way can be construed as granting rights to persons other than the contracting parties. Although the courts may quibble, the parties have certainly made their intention clear.

Why is this defensive tactic necessary? Simply because it has been long recognized that a third party can sue as a beneficiary on a contract that accrues to his or her benefit. In order for a third person to be considered a beneficiary under a contract, that person must prove the intent to benefit a third person. The benefit must not be merely incidental, but immediate and to such a degree as to indicate the assumption of a duty to make reparation if the benefit is lost. Absent an intent to benefit, the third person is merely an incidental beneficiary, one with no right to enforce the contract. An incidental beneficiary may derive a benefit from the contract even though he or she is not a contracting party. But the incidental beneficiary does not have the right to enforce the contract.

The concept seems strange. You and another establish a contract, yet a third person can enforce it. Just how did this situation arise? The milestone case in New York is *Lawrence v. Fox*, 20 N.Y. 268 (1859). The facts in that case are simple enough. Holly loaned Fox $300. At the time he loaned Fox the money, Holly stated that he owed that amount to Lawrence and had agreed to pay it to him the next day. Fox, receiving the money, promised to pay it to Lawrence the next day. The court affirmed the principle of law, long recognized and early established, that when one person engages with another by contract to do some act for the benefit of a third, that third person may maintain an action for breach of the contract. The court notes that such was the law of New York as early as 1806 and finds for Lawrence. While 1806 may be the earliest case in New York, the court in *Pound v. New Rochelle Water Co.*, discussed below, notes that the leading case in England is *Dutton v. Pool*, 1 Ventris 318, which was decided in the seventeenth century during the reign of Charles II.

What sort of contracts have been examined in determining whether or not third-party liability exists? *Fieger v. Glen Oaks Village, Inc.*, 309 N.Y. 527 (1956), involved a contract between Glen Oaks Village, the owner and developer of a group of 11 apartment houses and the Federal Housing Authority. Fieger was a tenant. Fieger complained that, because of fraud and gross neglect by Glen Oaks Village, the cost of constructing the apartment buildings was overestimated as a basis for obtaining from the Federal Housing Authority $24 million in mortgages. It was Fieger's contention that Glen Oaks Village had constructed the development in an inferior manner. Glen Oaks Village had submitted applications for mortgages on the 11 buildings. It had estimated total costs of construction, according to the drawings and specifications, at about $26.7 million. Glen Oaks Village requested, and the Federal Housing Authority (FHA) granted, insurance for mortgages on the 11 buildings aggregating $24 million. Pursuant to the FHA requirements, Glen Oaks Village filed rental schedules with the FHA administrator. The rents were fixed at an amount to cover operating expenses, mortgage payments, taxes, and a reasonable return to Glen Oaks Village. The crux of Fieger's complaint was that the maximum rent established bore a direct relationship to the amounts of the mortgages.

Fieger contended that the FHA approved maximum rents based on the figures, but that the total cost of constructing the buildings was not $26.7 million, but only $20 million. The $24 million in mortgages was about $4 million in excess of the actual cost of the buildings and incidentals. Fieger complained that this $4 million was distributed to Glen Oaks Village and that the annual rentals paid by all tenants in the buildings were in excess of approximately $360,000 per year.

The court found two reasons why Fieger could not prevail. The excessive rentals were fixed by the FHA administrator. This was a federal governmental action by authorized federal officers. The state courts have no power to review official acts performed by federal officials under the authority of acts of Congress. When the federal authorities and officers acted within their authority and fixed the rents, the courts of the state could not revise or review those determinations.

But that is not the issue we are concerned with. Fieger argued his status as a third-party beneficiary. The doctrine allows a beneficiary of a contract to enforce that contract, even though that beneficiary is not a contracting party. The court considered the contract between Glen Oaks Village and the FHA. The contract is made for the benefit of a large class of American citizens, including Fieger, who, at the time, could not find adequate housing. The primary design of the National Housing Act was to stimulate building, increase employment, and benefit low-income tenants by providing them with suitable housing accommodations at reasonable rentals. But the making of this contract was an act of government and, as all such acts are supposed to be, for the benefit of some or all of the citizens. By statute, the FHA administrator was to approve the rent schedule not only to prevent the charging of unconscionably high rents, but also to ensure that the rents were adequate to cover operational expenses, including mortgage payments. Even if the court had distorted this arrangement into a simple contract for the benefit of tenants, it would still have been true that Fieger was not actually suing as a third-party beneficiary to enforce the contract between Glen Oaks Village and the FHA. Rather, Fieger and the other tenants were attempting to bypass that contract, since the rents they paid were rents fixed by the FHA administrator pursuant to that contract. The only contracts that Fieger and the other tenants were parties to were their own leases with Glen Oaks Village. The rents they were paying were fixed in those leases. The tenants alleged not that Glen Oaks Village set out to harm them, but that Glen Oaks defrauded the United States government into fixing rents that would not have been so fixed had Glen Oaks Village not defrauded the government. The court found that the tenants' complaints could not be remedied unless some legislative body, such as the Congress, provided them with a remedy. The tenants were not third-party beneficiaries of Glen Oaks' contract with the FHA. The court found for Glen Oaks Village.

Gordon v. Holt, 65 A.D.2d 344, 412 N.Y.S.2d 534 (4th Dept. 1979), appeal denied 47 N.Y.2d 710, 419 N.Y.S.2d 1026 (1979), involved the actions of a public body. Gordon was one of the purchasers of an apartment complex, Southwind Terrace. Within three months after purchase, Southwind Terrace was destroyed by fire. The litigation that ensued involved the present owners; the former owners; the builder; the design architect; the supervising architect; the City of Utica, New York; and two of its employees. During construction of Southwind Terrace, the insurance company providing the mortgage engaged a local architectural firm to make periodic inspections and to report on the progress and quality of the construction. The supervising architectural firm indicated that the construction was of good quality and conformed substantially with the plans and specifications. Gordon and his companion prospective purchasers became interested in purchasing Southwind Terrace. As a condition of the sale, the initial owner, Georgian Courts, was required to furnish an updated certificate of occupancy. The City of Utica issued a new certificate of occupancy, even though the Utica building inspector discovered major violations of the building code. Gordon and the insurance company for several of the tenants brought the supervising architectural firm into the lawsuit, arguing that the firm had wrongfully reported to the insurance company providing the mortgage that Southwind Terrace was properly constructed. The court found for the supervising architectural firm.

The firm had been engaged by Metropolitan Life, the insurance company providing the mortgage, to provide certain services. There was no contractual relationship between the architectural firm and either Gordon or the tenants. The court noted that there are certain instances when a party to a contract may become liable to noncontracting third parties, but indicated that this was not such a case. The architectural firm was employed by Metropolitan to protect Metropolitan's security interest in the building and it owed no duty to an indeterminate class of persons such as future owners, Gordon, or tenants. The court found no arguable relationship between the architectural firm and any tenant in Southwind Terrace, nor was there a relationship to any subsequent purchaser of Southwind Terrace. Therefore, the court found no duty owed by the supervising architects arising from reports submitted exclusively to Metropolitan.

Claims were also asserted against the City of Utica and its two employees based on the admitted fact that the City issued a certificate of occupancy for a building that did not comply with the provisions of the building code. The court found against the City and its employees.

It was clear that the initial owner, Georgian Court, had requested the City to issue a new certificate of occupancy. In spite of the inspector's discovery of major violations of the building code, including the failure to provide fire walls, that certificate was issued. Gordon relied on the truth of the certificate in deciding to purchase Southwind Terrace. The court noted that, although the City was under no duty to issue the certificate merely because Georgian Court had requested it, once the City voluntarily undertook to inspect with a view to issuing a new certificate of occupancy, it was required to act with reasonable care. The court noted that a trial could reveal that Gordon did not, in fact, rely on the certificate, that the reliance was not reasonable, that his own negligence contributed to the loss, or that the lack of fire walls was not the proximate cause of the loss. However, at this juncture, it is too early to deem the City not liable to Gordon.

The tenants were also suing the City. The court noted that it was first necessary to have found that City owed a duty to the tenants. Without a duty owed, there could be no breach from which liability would flow. The contention by the tenants that the City negligently failed to enforce its building and fire code was simply an assertion that the City failed to exercise its police powers to provide adequate protection to the tenants of Southwind Terrace. The court noted that the duty to provide adequate police and fire protection was owed to the public at large and is not actionable by a single individual, unless that individual could show that a municipality owed him or her a special duty. In contradistinction to the prospective purchasers, the court noted that the special duty may be created either by statute or by a course of action, which the municipality is not required to take but voluntarily undertakes to perform. In so doing, it would induce the individual to rely on the municipality's duty to exercise reasonable care in its performance. There was no special duty created by statute. Nor could the tenants point to a special relationship. The tenants did not allege that they detrimentally relied on any act by the City. Unlike the special relationship, which could be established by Gordon, there was no voluntary act by the City on which the tenants relied. The court found for the City.

Pond v. New Rochelle Water Co., 183 N.Y. 330 (1906), involved a lawsuit against a corporation. The water company, in exchange for the right to lay and maintain its water mains through the streets of the village for a number of years, contracted with the village to supply the inhabitants of the village with pure and wholesome water at an annual rate not exceeding a specified amount. Subsequently, the water company informed the village residents that, in the future, no water would be supplied except through water meters. The rates were clearly in excess of those specified in the contract with the village. Charles Pond, a village resident, sued, claiming to be a beneficiary of the contract between the village and the water company. The court noted that the village sought to protect its inhabitants who, at the time of the execution of the contract, were consumers of water and those who might thereafter become so, from extortionate rates imposed by a water company. It granted that company a valuable franchise extending over a long period of time. As such, the court found that Mr. Pond is an intended beneficiary with the right to enforce the contract.

Eastern States Elec. Contractors, Inc. v. William L. Crow Const. Co., 153 A.D.2d 522, 544 N.Y.S.2d 600 (1st Dept. 1989), involved a subcontractor and a hospital. Eastern States had subcontracted with Crow Construction to perform electrical renovation work on a building owned by the Methodist Hospital. The subcontract between Eastern States and Crow specifically incorporated a clause excluding any contractual liability on the part of Methodist Hospital. The court noted that generally, a subcontractor may not assert a claim against an owner with whom it is not in contract. This, joined with the exclusory clause in the contract, allowed the court to release the hospital from the lawsuit.

James McKinney & Sons, Inc. v. Lake Placid 1980 Olympic Games, Inc., 92 A.D.2d 991, 461 N.Y.S.2d 483 (3rd Dept. 1983), aff'd as modified 61 N.Y.2d 836, 473 N.Y.S.2d 960 (1984), involved a construction project manager. McKinney entered into a contract with the Lake Placid 1980 Olympic Games to fabricate and erect certain steel structures for use in the construction of the Olympic field house. Gilbane Building was the project manager, Kahn & Jacobs the project architect, and Gillum Associates the design engineer. Although McKinney instituted suit against others as well, these three entities are our concern here. The court noted that there was no contractual relationship between Gilbane, the project manager, and McKinney. Therefore, the action for breach of contract must fail. In addition, the court rejected McKinney's attempt to claim negligence on the part of the architects who had contracted to provide services to the owner. By the same reason, the court found no liability on the part of the design engineer, Gillum. Since there was no contractual relationship or any independent duty between McKinney and Kahn, the project architect, or McKinney and Gillum, the project engineer, McKinney had no claim against those two entities.

Of concern to us in *Martirano Const. Corp. v. Briar Contracting Corp.*, 104 A.D.2d 1028, 481 N.Y.S.2d 105 (2nd Dept. 1984) is the complaint by Martirano, a subcontractor, against a landowner and his agent. The court found that since the landowner and agent were not in privity with the subcontractor, such an action could not be maintained.

Briar Contracting had undertaken to perform all the work and furnish all the material and equipment necessary for the construction of the Edith Macy Conference Center for the Girl Scouts. Tishman Construction acted as the agent for the Girl Scouts and contracted to pay Briar in accordance with the terms of the agreement. Briar entered into a subcontract with Martirano, who agreed to perform all masonry work as outlined in the contract between Tishman and Briar. During the course of the project, numerous disputes arose between Martirano and Briar concerning the progress and quality of Martirano's work. Both Martirano and Briar claimed that the other had breached the subcontract agreement.

Martirano's claims against Tishman and the Girl Scouts concern us here. The court noted the general rule that a subcontractor may not assert a claim that is contractual against persons with whom it had no contract. There exists no liability to incidental beneficiaries not in contract. The court noted no indication that Tishman or the Girl Scouts had contracted with Martirano. A breach of contract claim by Martirano against Tishman or the Girl Scouts was disallowed by the court. No contract existed between Martirano and either of those entities. In addition, Martirano cannot be considered an intended beneficiary of the contract between Tishman as agent for the Girl Scouts and Briar.

§ 7.1.2 Basis of Rule

It has often been said that a person claiming to be a third-party beneficiary has the burden of demonstrating an enforceable right. Is there a uniform theoretical basis for this third-party beneficiary rule? Hardly. A review of the cases reveals that the courts look hard at the facts presented.

As a contractor, what is your best defense? You are willing to accept the problems you may have with the owner or with your subcontractor. You need not assume any liability to third persons. Your contract should clearly state that no rights accrue to third persons as a consequence of the contract.

§ 7.1.3 Made and Intended for Benefit of Third Person

Under what circumstances will the court deem a contract to have been intended for the benefit of a third person? *Majestic Mfg. Corp. v. L. Riso & Sons Bldg. Co.*, 27 N.Y.S.2d 845 (Queens Co. 1940), aff'd 261 A.D. 1099, 27 N.Y.S.2d 846 (2nd Dept. 1941), involved the construction of a factory for Majestic. Riso, as general contractor, had agreed to construct the factory for Majestic and had entered into a subcontract with Kerstein and Zaville for the performance of certain plumbing work in the factory. Majestic sued both Riso on his contract and Kerstein and Zaville claiming that it was a third-party beneficiary on the subcontract. The decision for this court was simple. It found no contract between Majestic and Kerstein and Zaville. Furthermore, a provision of the contract between Majestic and Riso provided that Riso was fully responsible for the acts and omissions of his subcontractors and that nothing in the contract shall create any contractual relationship between any subcontractor and the owner, Majestic. The court found nothing in the subcontract for the plumbing work which required performance for Majestic. Therefore, the court found that Majestic had no complaint against Kerstein and Zaville.

Hanssel v. P. Tomasetti Contracting Corp., 8 N.Y.S.2d 873 (Eire Co. 1939), aff'd 257 A.D. 1031, 13 N.Y.S.2d 565 (4th Dept. 1939), rev'd on other grounds 283 N.Y. 164 (1940), reh'g denied 283 N.Y. 697 (1940), involved the construction of a sewer project for the Buffalo sewer authority. The general contractor, Tomasetti, subcontracted with Hanssel for the construction of the Cornelius Creek intersecting and overflow chamber. Hanssel brought an action against both Tomasetti and the sewer authority. The court found that Hanssel had no valid claim against the sewer authority and dismissed Hanssel's complaint as against the authority. It noted that there was no contract between Hanssel and the authority. The contract between Hanssel and Tomasetti was an independent contract not made or signed by sewer authority, nor made for its benefit. The general contract, made by Tomasetti and the sewer authority was not made or signed by Hanssel, nor for his benefit. Hanssel is simply not a beneficiary under that contract, nor obligated in any manner by its provisions, save by a reference in the general contract to the subcontract between Hanssel and Tomasetti.

The general rule is that there must be a relationship between the parties for the maintenance of an action on contract. If the contract could not be enforced against Hanssel, it cannot be enforced by him. The exception to the general rule is when the contract is made for the benefit of a third person. The court noted that it was insufficient that the performance of the contract may benefit the third person. For a third person to have the right to enforce a contract, it must have been entered into for his or her benefit, or such a benefit must be the direct result of the performance of the contract and within the contemplation of the contracting parties. The court found that Hanssel did not come within the exception. Hanssel had no legal or equitable claim for the benefit of the contract. The court noted the ingenious argument that under the National Recovery Act, all municipal improvements in which the federal government provides assistance are for the interest of all classes and individuals. The court stated that the loaning or even giving away of federal funds for municipal improvement does not yield to every corporation, firm, and individual a tangible community of interest in the profits of the fulfillment of such improvement. The court is lead to the irresistible conclusion that Hanssel has no claim against the authority.

A different set of facts presents itself in *Vance v. Yonkers Contracting Co.*, 280 A.D. 839, 113 N.Y.S.2d 733 (2nd Dept. 1952). Mr. Vance was a property owner and was suing Yonkers Contracting, an excavation contractor on adjacent property for injury to his real property caused by excavation and blasting. In a contract with the owner of the property being developed, Yonkers Contracting agreed that it would protect the owners of adjacent property from injuries arising from the carrying out of the contract and make good any damage or injury. The court found that the contract provision between the owner and Yonkers Contracting was made directly for the benefit of a class to which Vance belonged. Vance was clearly a third-party beneficiary and the court found for Vance.

The fact that the third party is not identified, or even identifiable, when the contract is made does not prevent that party from enforcing the contract. However, the third party has no right to enforce the contract until such time as he is identified. The general rule is that a person claiming to be a third-party beneficiary has the burden of proving that he or she has an enforceable right.

In *Bonwell v. Stone*, 128 A.D.2d 1013, 513 N.Y.S.2d 547 (3rd Dept. 1987), Stone had installed a septic system for a home at the request of a realtor. The Bonwells had contracted to purchase the house. The contract was contingent upon a building inspection. The inspection revealed that the home septic system was defective. Subsequently, the sellers' realtor contacted Stone and arranged to have the system repaired. Stone made the repairs, and its bill was submitted to and paid for by the realtor.

After the Bonwells took possession of the home, they discovered that the septic system was not functioning properly. They ultimately employed another contractor to perform extensive repair work. They sought to be reimbursed from Stone.

Stone argued that his contract was with the realtor. The realtor was the sellers' agent. The Bonwells were simply not a party to the contract. The court reviewed the general rule, a party seeking to recover as a third-party beneficiary under a contract must establish that 1) a valid and binding contract exists between the other parties; 2) the contract was intended for his or her benefit; and 3) the benefit is immediate rather than incidental. The court cautioned that a person who could not show the intent of the contracting parties to benefit him or her simply has no right to recover, since he or she is merely an incidental beneficiary. The court noted that Stone's dealings were with a realtor. This compelled the inference that Stone must have known that the house was to be sold and that the buyer would benefit from the repairs. The court found for the Bonwells.

A third-party beneficiary can present proof that he or she was intended to be benefited even if the contract is silent as to including or excluding third-party beneficiaries from its coverage. The role of the court is, of course, to determine the weight to be given to that evidentiary proof in conjunction with the absence or inclusion of a third-party beneficiary provision in the contract. *Snyder Plumbing & Heating Corp. v. Purcell*, 9 A.D.2d 505, 195 N.Y.S.2d 780 (1st Dept. 1960), involved a dispute between Snyder, a plumbing contractor, and a general contractor in connection with the construction of a county home and infirmary. There were multiple prime contracts. Snyder had entered into a contract with the county to perform the plumbing and drainage work at the home and infirmary. Purcell, under a separate contract, was to perform the general contracting work. The contracts were similar except for specifications regarding the work to be performed by each contractor. Snyder claimed to be a third-party beneficiary of the contract between Purcell and the county, relying on an article that provided that should another (here, Snyder) suffer through acts or negligence of the general contractor and assert a claim against the country, then the county would notify the general contractor, who would pay any judgment or claim to be found against the county.

The court briefly reviewed Snyder's litigious enterprises before the initiation of this suit. Snyder had previously commenced an action against the county for damages attributable to delay caused by the county in the erection of the home and infirmary. The county ultimately paid Snyder, and Snyder delivered a general release to the county. The release reserved Snyder's rights, which it might have against other contractors involved in the construction. Thereafter, Snyder demanded arbitration against Purcell. Arbitration was denied on the grounds that there was no contractual relationship between Snyder and Purcell. At this juncture, Snyder commenced the lawsuit to recover damages for losses it sustained due to Purcell's procrastination in completing his work as general contractor. The amount of the lawsuit was precisely the amount that it had claimed and recovered against the county.

Snyder contended that the provisions of the contract between Purcell and the county gave Snyder the right to proceed against Purcell as a third-party beneficiary. The court found that the provision was solely for the benefit of the county and did not create an obligation for other contractors performing work. The county had never claimed that Snyder's loss was caused by Purcell. After asserting its claim against the county, Snyder had settled and released the county. The county had never demanded that it be reimbursed by Purcell, pursuant to the contractual clause. The court found that this proved that there is no derivative liability to be asserted against Purcell. Nor was there a duty by Purcell to Snyder. Snyder's reservation of rights against Purcell in the release he delivered to the county could not enlarge or alter the scope of the contractual provision. There was no duty to make Snyder whole pursuant to Purcell's contract with the county. The court found no such intent under the contract and held that Snyder could not maintain an action against Purcell.

To what lengths will persons go to attempt to show that they are third-party beneficiaries? *Wolfkeil v. Jamoco Air Conditioning Corp.*, 19 Misc.2d 655, 190 N.Y.S.2d 24 (Queens Co. 1959), involved the heating and air conditioning system installed in the Wolfkeil's house, which was manufactured by Westinghouse Electric. The Wolfkeils stated that Westinghouse's literature had induced them to purchase it. Westinghouse's brochure had stated, "... you are assured of Westinghouse *undivided responsibility* for dependable operation." The Wolfkeils claimed that this advertisement was a contract made for the benefit of a third party. The court clearly noted the absence of a contract between Westinghouse and someone else for the benefit of the Wolfkeils. Therefore, third-party rule had no application. The general advertising of Westinghouse was not a contract. The court found for Westinghouse and released it from the lawsuit.

Vandewater & Lapp v. Sack Builders, Inc., 20 Misc.2d 677, 186 N.Y.S.2d 103 (App. T. 1959), involved a dispute between engineers and surveyors and land owners. The engineers and surveyors had performed certain work for correcting drainage maps. The land owners, Sack Builders, refused to pay, claiming that previous maps that were prepared by Vandewater & Lapp for a prior owner were incorrect. Vandewater & Lapp had prepared a plot plan and a drainage map for Weinstock. The map was certified and approved by various departments and filed with the County Clerk. The property was subsequently sold to a realty company who, in turn, sold the property to Sack Builders. Construction was commenced on the property by Sack Builders in accordance with the elevations and grades certified by Vandewater & Lapp. During the course of construction, it became apparent that if the grades and elevations were followed, a pre-existing house would be subject to the flow of all runoff water onto its premises. The owner of that house approached various authorities. Those authorities rescinded the prior approval of the map and directed Sack Builders to prepare and file a new drainage plan. Sack hired Vandewater & Lapp to prepare the map, and Vandewater & Lapp did prepare corrected maps. Vandewater & Lapp sought compensation for the value of the work in preparing the corrected maps. Sack Builders claimed that the damages had resulted from the errors in the previous maps. Sack had expended monies in connection with those errors.

It was Sack Builders' contention that the original contract between Vandewater & Lapp and Weinstock entitled them to be considered a third-party beneficiary. Vandewater & Lapp argued that, since their prior contract was with Sack Builders, no liability could attach by reason of the erroneous maps. The court found that the original plot plan and drainage map had the effect of establishing Sack Builders as a member of a class intended to be benefited by the contract between Vandewater & Lapp and Weinstock.

That being the case, Sack Builders was entitled to be recovered for the damages it suffered as an intended beneficiary of the contract between Vandewater & Lapp and Weinstock.

In *Martin v. Dayton Seaside Corp.*, 25 Misc.2d 264, 205 N.Y.S.2d 573 (Queens Co. 1960), Edward Martin, a taxpayer, sought to stop Dayton from using portions of property acquired by it pursuant to the Federal Housing Act as public parking areas. Pursuant to a contract between the City of New York and Dayton, Dayton was not to use, rent, license, or permit temporary use of buildings, structures, or vacant land except with temporary arrangements with occupants situated on the land on the date of acquisition of title and for temporary relocation of site occupants to deferred areas. During the summer months, the land was used for parking purposes. That use was with the approval and sanction of various city and federal agencies. The use was conditioned upon Dayton's compliance with certain regulations. Charges for parking were fixed in accordance with a schedule set by a city agency. All costs and expenses in the preparation of the area for parking were met by Dayton and the net proceeds of the operation were deposited in a project account, subject to audit by city and federal authorities. Martin claimed he had a right to sue as a third-party beneficiary. The court pointed out, however, that the language of the contract between Dayton and the City provided that nothing in the contract shall be intended or construed to give any person, firm, or corporation other than the parties hereto, the holder or holders of the first and second mortgages, and the Federal Housing Commissioner any right or claim under the agreement. The agreement was to be for the sole and exclusive benefit of the parties to the agreement, to the holders of the mortgages, and the Federal Housing Commissioner. The court consequently found for Dayton, since Martin had no right as a third-party beneficiary.

When are you the intended beneficiary? *Astor Terrace Condominium v. Schuman, Lichtenstein, Claman & Efron*, 583 N.Y.S.2d 398 (1st Dept. 1992), involved a dispute between condominium unit purchasers and engineering and design professionals. Engineering and design professionals had contracted with the sponsor of the condominium project. Individual condominium unit purchasers claimed various damages resulting from defects in the construction and design of the building.

The court noted that documentation, including the selling documents and the Offering Plan demonstrated the sponsor's intention to make the unit owners intended beneficiaries of the design contracts. This was discernable from the facts as well as the circumstances surrounding the contracts. Since design and construction failures affect the condominium unit owners' standard of living and their ability to sell, it cannot be said that the condominium owners were incidental, rather than intended, beneficiaries of the engineering and design contracts.

In addition, the design and engineering professionals must have been aware that the substance of their reports would be distributed to and relied on by the prospective condominium unit purchasers.

The reliance did occur and the conduct of the professionals linked them to the unit purchasers.

The court noted that the units were marketed as luxury condominiums with an emphasis by the sponsor that it had gathered the best engineers and architects to design and construct the building. The court found for the condominium purchasers.

§ 7.1.4 Incidental Benefit

If a person benefits incidentally from your contract, he or she will not be able to sue on it. In general, a construction contract will not afford benefits to persons other than the contracting parties. Unless the contract contains an express provision to establish others as intended beneficiaries, the courts generally will consider such persons as incidental beneficiaries.

As we have often seen, the determination as to whether a beneficiary is intended and thereby capable of suing on a contract, or merely incidental with no rights on that contract, is a determination that is based squarely on the facts. In *Fourth Ocean Putnam Corp. v. Interstate Wrecking Co.*, 66 N.Y.2d 38, 495 N.Y.S.2d 1 (1985), a dispute arose between Fourth Ocean, the owner of certain property, and the wrecking company that demolished that property. Fourth Ocean had owned the Atlantic Beach Hotel. The Hotel was severely damaged in a fire. The village, acting properly under a village ordinance, obtained a court order to compel Fourth Ocean to demolish the structure as a public nuisance and a dangerous and unsafe fire hazard. Fourth Ocean failed to demolish the structure. The village entered into a demolition contract with Interstate as authorized by the court order.

Interstate was to demolish and remove the building and boardwalk. All materials and debris were to be removed from the premises. After demolition, the lot was to be compacted and graded to "grade level" with dirt. Walls and foundations were to be removed one foot below the elevation of existing grade. Slabs were to be crushed for drainage and large pieces removed. Before demolition commenced, Fourth Ocean attempted to obtain an injunction to prevent it. It failed. Demolition was commenced and concluded. The costs of the demolition were subsequently recovered from Fourth Ocean by the village.

Some years later, Fourth Ocean began to construct new buildings on the site. During initial excavation, it discovered walls and foundations had not been removed one foot below existing grade, nor had the slab been crushed and large pieces removed as required by the contract. Fourth Ocean moved against Interstate, claiming it was a third-party beneficiary of the contract between Interstate and the village.

The demolition performed by Interstate was not to prepare Fourth Ocean's property for further income-producing development, but to benefit the village at large by removing a structurally dangerous and unsafe building. Fourth Ocean's recalcitrant position with respect to the demolition indicated that it was merely an incidental beneficiary without the right to maintain an action. The court noted that the absence of a duty to the beneficiary has been held to negate an intent to benefit in both governmental and private contract cases. The court emphasized that, when upholding a third party's right to enforce a contract, a consideration should be that no one other than the third party can recover if the contract is breached. Additionally, the contract should be reviewed for an indication evidencing an attempt to permit enforcement by a third party. The court concluded that Fourth Ocean was an incidental rather than an intended beneficiary.

Fourth Ocean had argued that because the work was performed to satisfy its obligation, the contracting parties must have intended to benefit it. However, the work was being performed not as a means of benefiting Fourth Ocean, but to remedy Fourth Ocean's default in order to protect the public against a public nuisance as authorized by ordinance and court order. Other than the specifications, which required that the walls and foundations be removed one foot below the elevation of existing grade and the slab crushed for drainage and large pieces removed, there was nothing in the contract, from the full performance of which Fourth Ocean might ultimately benefit. There was no suggestion of an intent to benefit Fourth Ocean. The contract provision with respect to the demolition was for the benefit of the village, which had a contingent interest in the performance arising from the possibility that, if Fourth Ocean failed to reimburse the village for the contract expenditures, the village would become the owner of the property through forfeiture. There was nothing that in any way suggested that the village or Interstate contemplated that Fourth Ocean would rely on this contract. The court found that Fourth Ocean was not an intended beneficiary.

Edward B. Fitzpatrick, Jr. Constr. Corp. v. County of Suffolk, 138 A.D.2d 446, 525 N.Y.S.2d 863 (2nd Dept. 1988), appeal denied in part, appeal dismissed in part 73 N.Y.2d 807, 537 N.Y.S.2d 477 (1988), reconsideration denied 73 N.Y.2d 918, 539 N.Y.S.2d 302 (1989), involved the termination of a construction company's contract with the county. Bowe Walsh had contracted with the county to provide certain engineering services in connection with the planning, design, and construction of a sewer pipeline for the Suffolk County Southwest Sewer District's water pollution control plant at Bergen Point. Subsequently, Bowe Walsh subcontracted with Converse Ward to provide a report outlining the results of certain ocean, bay, and land soil borings.

Fitzpatrick was the successful bidder to construct a portion of the sewer pipeline in question. That pipeline was to run under the Great South Bay. Fitzpatrick purchased performance and material bonds. Subsequently, it became apparent that

Fitzpatrick would not be able to complete the construction of the pipeline within the time required by the contract. The county terminated Fitzpatrick's services. The County and Fitzpatrick's sureties then entered into an agreement whereby the sureties became the construction contractor. The agreement allowed the sureties to engage Fitzpatrick as a subcontractor to continue construction of the pipeline.

Fitzpatrick began this litigation and claimed it was a third-party beneficiary of the various contracts between the county, Bowe Walsh, and Converse Ward. The court noted however, that two of the contracts contained clauses to the effect that no third-party rights accrued therefrom. It is a general rule that, where a provision in the contract expressly negates enforcement by third parties, that provision is controlling. The one remaining contract, a contract between the County and Converse Ward, contained a provision that Converse Ward's services would run directly to the county. The court's interpretation of this language was that Fitzgerald was merely an incidental beneficiary of the services performed by Converse Ward. As an incidental beneficiary, Fitzpatrick could not claim third-party beneficiary status.

§ 7.1.5 Identification of Third Person

The identification of the third person speaks to the question of whether or not that person will be allowed to sue on the contract. Where the court finds that the facts and the contract provisions indicate an intention to benefit a third person, then that person falls into the classification of an intended beneficiary who may enforce the contract. *Bethpage Water Dist. v. Hendrickson Bros., Inc.*, 138 A.D.2d 660, 526 N.Y.S.2d 476 (2nd Dept. 1988), involved a dispute between the water district and Hendrickson regarding 24 water main breaks. Hendrickson contracted with the county to install certain sewer improvements below the water district's mains. The water district claimed it was a third-party beneficiary of the sewer installation contract between Hendrickson and the county. The court looked to the express terms of that contract. Hendrickson was obligated to protect, in a suitable manner, all utilities encountered and to repair any damages to structures, utilities, and facilities caused by Hendrickson's operations. The court found that this provision unambiguously evidenced the contract parties' intention to insure uninterrupted utilities service. There was a corresponding intention to benefit the owner of the utility by imposing on the sewer contractor the burden of protecting and repairing utility installations. The court found for the water district.

In *Charlebois v. J.M. Weller Associates, Inc.*, 136 A.D.2d 214, 526 N.Y.S.2d 648 (3rd Dept. 1988), aff'd 72 N.Y.2d 587, 535 N.Y.S.2d 356 (1988), the Charleboises contracted with Weller for a new warehouse and addition to an existing building from which they operated their beer distributorship business. The contract denominated a construction team consisting of the contractor, Weller Associates, the owners, Charleboises, and an architect engineer, James M. Weller, P.E. Construction began but disputes arose between the parties over cost, design, building code compliance, and other alleged defects. The Charleboises refused to make further payments, leaving approximately $600,000 in construction costs owed until all disputes were resolved to their satisfaction.

This was a traditional "design-build" contract. Weller Associates agreed to be responsible for furnishing design and requiring that all architectural and engineering services be provided by a licensed, professional engineer. Weller Associates, the contractor, could not provide those engineering services, only a licensed professional could. Therefore, Weller Associates subcontracted with James M. Weller, P.E., for his professional engineering expertise. The Charleboises argued that they were a third-party beneficiary of that subcontract between Weller Associates and Mr. Weller. The contract between the Charleboises and Weller Associates specified a named, licensed, professional engineer to perform all services for which a license was required. The court found that the Charleboises were third-party beneficiaries of the subcontract between Weller Associates and the licensed professional engineer.

Arrow Louver and Damper Div. of Arrow United Industries, Inc. v. N.Y.C. Transit Authority, 106 A.D.2d 533, 482 N.Y.S.2d 844 (2nd Dept. 1984), involved a number of contracts entered into by the Transit Authority, acting on behalf of both the City of New York and the Metropolitan Transportation Authority. Those contracts provided for, among other things, the furnishing and installation of damper equipment at various subway stations. Arrow Louver was a manufacture of dampers. Arrow contended that the various contracts set forth certain definite standards for dampers and that its dampers met those standards.

The court observed that, in general, a subcontractor is not a third-party beneficiary of a contract between an owner and a prime contractor, but rather, merely an incidental beneficiary. There was no allegation that Arrow was intended by the contracting parties to be third-party beneficiary. Arrow's contention that the contracts explicitly referred to the Arrow damper was insufficient, since the same contracts provided that the reference to a particular manufacturer was meant only to provide an example for the purpose of indicating requirements of type, quality, and design. The court found no facts that would warrant a finding that Arrow was a third-party beneficiary.

As a stranger to the contracts, Arrow lacked standing to sue for enforcement of the contractual provisions.

Castorino v. Unifast Building Products Corp., 161 A.D.2d 421, 555 N.Y.S.2d 350 (1st Dept. 1990), is perhaps the most egregious example of stretching for third-party beneficiary status. DCI was general contractor for the owner of an apartment building. DCI executed a subcontract with Unifast to supply and install windows throughout an apartment building. It was contended that a woman murdered in the apartment building was an intended third-party beneficiary of the subcontract between DCI and Unifast.

The court once again notes that an intended beneficiary may maintain an action as a third party. To do so, that third party must establish: 1) the existence of a valid and binding contract between other parties; 2) that the contract was intended for her benefit; and 3) that the benefit is sufficiently immediate, rather than incidental, to indicate the assumption by the contracting parties of a duty to compensate the third party if the benefit is lost.

The court found nothing in the subcontract between Unifast and DCI that evidenced a discernable intent to allow recovery for the specific damages that the third party suffered. The theory was that an assailant had murdered the decedent by gaining entry either through a window that did not have locking devices or that did not have locking devices in proper working order. Since the windows could not be properly closed or locked, there was an attempt to hold Unifast liable. The court found that the deceased tenant was not an intended beneficiary of the subcontract between DCI and Unifast and, therefore, suit could not be maintained on a third-party beneficiary theory.

§ 7.1.6 Acceptance of Contract by Third Person

Is the willingness of a third person enough to create a beneficiary relationship? In *Wright v. Herb Wright Stucco, Inc.*, 72 A.D.2d 959, 422 N.Y.S.2d 253 (4th Dept. 1979), rev'd. 50 N.Y.2d 837, 430 N.Y.S.2d 52 (1980), Raymond Wright was an employee of Wright Stucco. Jewel Builders was the general contractor for a residential project being constructed by a subsidiary of the New York State Urban Development Corporation. Jewel Builders was required to pay prevailing wages and to impose that obligation on each of its subcontractors. Mr. Wright was paid less than the prevailing wage and sought to enforce his rights as a third-party beneficiary of the contract between Jewel Builders and the subsidiary of the Urban Development Corporation.

It had long been the policy of New York State that workers on public projects were to be paid according to the prevailing wage in the locality. The court noted that the labor law had, as its entire aim, the protection of working men against being induced or obligated to accept wages below the prevailing rate and must be construed liberally to carry out its beneficent purposes. The court allowed Raymond Wright to proceed upon the third-party beneficiary principle.

Gjertsen v. Mawson & Mawson, Inc., 135 A.D.2d 799, 522 N.Y.S.2d 891 (2nd Dept. 1987), involved Finn Gjertsen who was a dock worker at a construction site on the East River. He was injured while unloading steel plates. He contended that the means to unload those plates were inadequate. The labor law specifically requires that all contractors are to provide proper equipment in such circumstances.

Sand was the general contractor on the site. Ingram & Greene was Sand's subcontractor. Ingram & Greene entered into a sub-subcontract with Rogers, Inc., who employed Gjertsen.

Gjertsen's employer, Rogers, Inc., paid him workman's compensation. Gjertsen commenced suit for his injuries against the general contractor, Sand and the subcontractor, Ingram & Greene. There was a direct contract between Rogers, Inc.

and Ingram & Greene. H. Sand, the general contractor, sought to be indemnified by Rogers, Inc. as an intended third-party beneficiary of the Rogers, Inc. and Ingram & Greene subcontract.

As general contractor, Sand was clearly an intended beneficiary of the sub-subcontract between Ingram & Greene and Rogers. Therefore, Sand could maintain a claim for indemnification against Rogers, Inc.

Individual complainants were successful in the two prior cases. In *City of New York v. Consolidated Edison Co. of New York, Inc.*, 114 A.D.2d 217, 498 N.Y.S.2d 369 (1st Dept. 1986), the attempt of a corporation to assert third-party status was reviewed. A public utility such as Con Edison is required, at its own expense, to remove or protect its underground facilities whenever the public health, safety, or convenience requires that this be done. Con Edison operated pursuant to a franchise granted by the City. Con Edison allowed the City or its contractor a right-of-way through, under, or above any part of Con Edison structures now, or at some future time, placed in any street or public place in the City. If it became necessary for the City to construct any structure, Con Edison would, at its own expense protect, alter or relocate all, or any portion of its structures and equipment. Should Con Edison refuse to do so, the City had the right to break through, remove, or relocate any or all structures, or portions of structures and Con Edison would pay the City the expenses incurred.

The City had entered into a contract with Perez Interboro Asphalt for the reconstruction of a street. The construction project covered a distance of four to five miles. It involved excavation of the existing roadbed, installation of new water mains, installation of sewer drainage, catch basins and connecting chutes, regrading and repaving of the roadbed, installation of traffic and street lights, and the planting of trees. Soon after construction commenced, Perez encountered substantial interferences from Con Edison's subsurface facilities, electrical ducts, and duct banks. Con Edison had received timely statutory notice of commencement of the projects and became obligated to remove or project the interferences encountered by Perez. The court noted that on project such as these, Con Edison and the City's contractor would have negotiated between themselves as to which facilities must be removed, which protected in place, and which methods of construction the contractor might employ to eliminate interferences and minimize removal expenses. Con Edison had the option to use the City's contractor, to hire its own contractor, or to perform the work itself. Con Edison would not have had the option to leave interferences in place. Con Edison was required by its franchise, by statute, and by common law to remove or protect in place at its own expense.

Con Edison and Perez were unable to reach an agreement as to the cost of eliminating the interferences. The City attempted to mediate, but was unable to bring the parties to an accord. In order to minimize further delays, the City sent directives to Con Edison to eliminate or remove certain interfering facilities and to clear and protect others. Con Edison refused.

Con Edison commenced an action against the City and Perez. Con Edison contended it was a third-party beneficiary between the City and Perez. Con Edison pointed out that the City's contract with Perez contained a condition that Perez, while performing work, would insure continuing service for all underground facilities. Perez was to provide prompt and adequate support for every underground facility located in the work area. The court pointed out, however, that an explanatory note provided that the section would not affect the right of any governmental unit to require the operator of such private underground facility to move or relocate it. The court noted that the provisions did not confer upon Con Edison the status of a third-party beneficiary. At best, Con Edison was an incidental, as opposed to an intended beneficiary of the contract. The court found against Con Edison.

§ 7.1.7 Consideration Between Contracting Party and Third Person

Is there some consideration that passes between a party to the contract and a third person? If there is, the courts may find that that third person is indeed a third-party beneficiary. Without consideration, it is highly unlikely that the court will find the third person an intended third-party beneficiary.

Costa v. Callanan Road Improv. Co., 15 Misc.2d 198, 184 N.Y.S.2d 534 (Ulster Co. 1958), was an action by Peter Costa against Callanan on a road improvement contract with the State. Costa was the owner of property likely to be affected by the contract work. He argued that he was an intended third-party beneficiary. The court examined the contract between Callanan and the state and found a third-party beneficiary clause. The clause stated that all damage of whatever nature resulting from the performance of the work including omissions and supervisory acts of the state would be borne by the contractor. All work would be solely at his risk until it was finally inspected and accepted by the State. The clause excluded from the contractor's responsibility damages resulting from faulty design or from willful acts of the State's officials or employees. The provision continued that nothing in the provision or the contract would create or give to third parties any claims or right of action against the contractor or the State beyond that which may legally exist irrespective of the provision or contract.

The court noted that for a third-party beneficiary theory to succeed, there must be an intent on the part of the State to secure some benefit for the third person. Pursuant to the clause, Costa was not intended to be benefited. There was no obligation or duty existing from the State to Costa.

The court found no intention on the part of the State to preserve in Costa any contractual cause of action either against Callanan or the State. On the contrary, the clause completely negated any third-party rights. The court found for Callanan.

Lake Placid Club Attached Lodges v. Elizabethtown Builders, Inc., 131 A.D.2d 159, 521 N.Y.S.2d 165 (3rd Dept. 1987), was an action brought by Lake Placid Club, a joint venture composed of owners of residential condominium units in a four-building project. Lake Placid Club argued that its members were intended third-party beneficiaries of the construction contract between a developer, Lake Placid Company, and the builder, Elizabethtown Builders. The members of Lake Placid Club either purchased their condominiums units directly from the developer or were successors to the original purchasers. Lake Placid Club was formed solely to prosecute damage claims for structural defects and resulting damages with respect to the condominium units. Lake Placid Club asserted contract claims based upon the contracts between the developer and Elizabethtown Builders and the developer and the architectural firm of Wallace, McHarg.

The court noted that a third-party beneficiary must prove the intention of the contracting parties to benefit that third person. Either performance will satisfy a money debt obligation or the circumstances will indicate that the third person is to receive the benefit of the performance. The court noted that there were no facts establishing the existence of an obligation by the developer to pay money to any of Lake Placid Club's members. Likewise, Lake Placid Club had failed to submit any evidence from the contractual language or other circumstances manifesting a mutual intention to confer rights upon the ultimate owners of the units. There was no showing of an intent on the part of the developer to give the ultimate owners any benefits from the performances promised to the developer by the builder or promised to the developer by the architect. There was nothing in the record to suggest that the developer had in mind anything but a normal business motive to obtain a construction project of sufficient quality for ready marketability to potential customers. Such a motive would not form the basis from which to infer the requisite intent to bestow performance benefits upon the ultimate purchasers of the condominium units, let alone their successors.

It was this circumstance that explains why, ordinarily, construction contracts are not construed as conferring third-party beneficiary rights. The court noted that once all of the condominium units had been sold, the developer might not have a reason to sue for nonperformance or for cognizable damages. The developer would have had enforceable rights and remedies for nonperformance during the life of the contract before such sale and is liable for construction defects after such sale. However, this factor was insufficient to demonstrate the existence of an intended third-party beneficiary situation. The court found for the developer and for the architectural firm.

§ 7.1.8 Contracting Party's Debt to Third Person

The general rule, and in New York from its earliest third-party beneficiary case, *Lawrence v. Fox*, 20 N.Y. 268 (1859), is that where two parties enter into a contract and one agrees to pay the debt of the other to a third person, that third person can sue for recovery. After reading the cases analyzed, this should come as no surprise. It bears repeating because it is the basis upon which third-party beneficiary theory is grounded.

§ 7.1.9 Public Contracts

As in other contracts, the rights of third parties in public contracts are defined by the contract documents. *New York Pneumatic Service Co. v. P.T. Cox Contracting Co.*, 201 A.D. 33, 193 N.Y.S. 655 (1st Dept. 1922), aff'd 235 N.Y. 567 (1923), involved a contract between Cox and the City of New York to change the grade of a street. One of the terms of the contract required that Cox move, alter, readjust, or rebuild pneumatic tubes and mail tubes as were necessary for the reconstruction of all surface, subsurface, and overhead structures directly or indirectly affected or disturbed by Cox in the progress of construction. The tubes were to be returned to their useful, safe, and durable condition as they existed before the construction work was begun. Cox failed to re-lay certain of the Pneumatic Service Company's tubes. The pneumatic tube company undertook that work and carried it to completion. The court found for the pneumatic tube company. It was clearly an intended third-party beneficiary of the contract between the City and Cox.

§ 7.1.10 Defenses

The closing moral of this chapter should be: Make your contracts clear and exclude third persons from their benefits. *Nepco Forged Products, Inc. v. Consolidated Edison Co.*, 99 A.D.2d 508, 470 N.Y.S.2d 680 (2nd Dept. 1984), involved yet another third person seeking to be deemed an intended beneficiary of a contract.

The Commerce Labor Industry Corporation and Con Edison had entered into a contract regarding the supply of utilities to Commerce Labor's subtenants in the former Brooklyn Navy Yard. Marsel Mirror had expended funds for the purchase of air compressors. Marsel argued that it was entitled to reimbursement from Commerce Labor as an intended third-party beneficiary of the contract between Con Edison and Commerce Labor. The court found against Marsel Mirror citing the provision of the Commerce Labor contract with Con Edison, which stated that nothing in the agreement, expressed or implied, was intended to confer on any other person any rights or remedies in or by reason of the agreement.

Where performance is to be made directly to a third party, that party is generally deemed an intended beneficiary and entitled to enforce the agreement where there is at least a presumption that the contract was for the benefit of the third party. However, the best evidence of whether the contracting parties intend their contract to benefit third parties remains the language of the contract itself. Where a contract provision expressly negates an intent to permit enforcement by third parties, then that provision is decisive. If you do not want third parties to benefit, say so.

CITATIONS

CHAPTER 8

ILLEGALITY

8.1 GENERALLY

§ 8.1.1 llegality

Generally contracting parties can bind themselves to any provisions they wish. There are, however, two restrictions. The provisions must not be illegal and the provisions must not be a violation of public policy.

What do the courts consider illegality and how does it affect your contract? *Murray Walter, Inc. v. Sarkisian Bros., Inc.*, 107 A.D.2d 173, 486 N.Y.S.2d 396 (3rd Dept. 1985), involved a dispute among the shareholders in Murray Walter, Inc., a New York construction corporation. When originally incorporated, there were four shareholders, Murray Walter (100 shares), James Matthews (100 shares), and George and Sarkisian (100 shares jointly). The shares owned by Matthews and the Sarkisians were not issued and their interest not disclosed because each was engaged as an independent construction subcontractor dealing with other general contractors.

Some years later, the stock certificates representing the shares owned by the Sarkisians were issued at their request. At about the same time, disputes arose between Walter and the Sarkisians, which, according to Murray Walter, Inc., significantly impeded its ability to conduct its business. The Sarkisians refused to personally guarantee Murray Walter, Inc.'s performance bonds, rendering it impossible for the corporation to obtain surety company bonding. Ultimately, Walter and the Sarkisians began negotiations toward liquidation of the interest of the Sarkisians in the corporation. The Sarkisians were offered $400,000 for their interest. They demanded, however, an amount approaching $1 million. By this time, their disputes had erupted into litigation. Agreement was finally worked out whereby the corporation would redeem the shares owned by the Sarkisians for $800,000. All parties would treat the exchange for tax purposes as a payment of Sarkisians' profit share in a totally unrelated joint venture. Murray Walter, Inc. was to report it as an ordinary business expense deduction, and the Sarkisians report it as ordinary income. The Sarkisians further agreed to indemnify Murray Walter, Inc. for one-half of any amount disallowed as an expense by the federal and state taxing authorities up to $400,000. The corporation alleged that it was agreed and understood that the maximum net cost of the acquisition of the Sarkisians' shares would not in any event exceed the amount of $400,000.

The Sarkisians did not report the transaction on their tax return as agreed upon. The Internal Revenue Service began an investigation of the Murray Walter, Inc. tax return, which had included the claimed deduction. Ultimately, both federal and state taxing authorities disallowed the Murray Walter, Inc. deduction of the purchase price as a business expense and assessed plaintiff the full tax deficiency plus interest totaling more than $800,000. No other civil or criminal penalties were imposed. Upon its making installment payments on the assessment, Murray Walter, Inc. demanded reimbursement from the Sarkisians. The demand was rejected. Murray Walter, Inc. claimed the Sarkisians fraudulently induced it to enter into this agreement.

The Sarkisians opposed Murray Walter, Inc.'s suit with the defense of illegality.

It was not seriously disputed that the parties' mutual agreement to falsely report the transfer of $800,000 was a violation of federal tax law. That false report alone would not render the contract per se unenforceable on public policy grounds unless a specific statute expressly so provided. In the absence of such an express provision, the words of the statute must be interpreted and the purpose of the legislature weighed. The court found the underlying facts insufficient to void the agreement because of illegality.

The court inquired into the underlying transaction. Murray Walter, Inc. averred that it was unwilling to purchase the Sarkisians' shares for more than $400,000 and only agreed to pay $800,000 if it could obtain a tax benefit equivalent to the difference. That tax benefit was to be accomplished by deducting the full price as an ordinary business expense. In agreeing to this arrangement, the Sarkisians made two promises: 1) to report the receipt of $800,000 as ordinary income, thereby avoiding any inconsistency with the plaintiff's reporting of the transaction; and 2) to reimburse the plaintiff in the amount of any tax disallowance of the $800,000 up to the maximum of $400,000. The court found that the Sarkisians' promise to indemnify Murray Walter, Inc. was not, in and of itself, illegal; certainly not in the same sense as the promise to falsely report the transaction. The indemnification provision was inserted to insure that whatever the consequences of reporting the full price as a deductible expense, Murray Walter's net cost for redeeming the Sarkisians' stock would not exceed $400,000. The court found this was confirmed by the absence in the agreement of any promise to defend Murray Walter, Inc. in a tax proceeding, or to underwrite any penalties or fines.

This contract was not partially executed. Murray Walter, Inc. had fully performed its promised exchange by paying the Sarkisians' interest. By Murray Walter, Inc.'s evaluation, it had paid more than double that which was called for under the agreement between the parties.

Before a contract has been executed by part performance, there is no serious harm in denying enforcement because of illegality. But, after one of the contracting parties has performed, denial of his or her remedies because of illegality would seriously penalize him or her. What, then, is the solution? Restitution may be awarded or the contract may be enforced according to its terms. If one party has substantially performed, the court should consider the extent of public harm and the relative guilt of the parties. The court analyzed the contract as two tax-related promises. The first was that in consideration of the payment of $800,000, there was one illegal promise to falsely report the transaction. There was another promise, though legal, to indemnify the corporation should the taxing authorities disallow the deduction. Where there are two promises, one legal and the other illegal, enforcement of the legal promise is permitted.

The court noted that the Internal Revenue Service chose not to impose either criminal sanctions or civil penalties. The court felt that the Sarkisians took unfair advantage of their superior bargaining position in view of the admitted ill health of Walter and imposed an inflated redemption price. The court noted that Walter believed that the tax benefits Murray Walter, Inc. would derive from reporting the payment as a business expense would be offset by the Sarkisians' tax burden for reporting it as ordinary income. The court performed a weighing. If the court refused to enforce the indemnification clause it would work a substantial forfeiture upon Murray Walter, Inc. while permitting the Sarkisians, who are at least equally guilty, to retain the entire financial benefit of the transaction. In order to avoid such results, courts have severed lawful from unlawful obligations of a contract for enforcement purposes. The overriding general policy of the courts is, as Justice Holmes once put it, "preventing people from getting other people's property for nothing when they purport to be buying it." The court held for Murray Walter, Inc.

§ 8.1.2 Presumptions

The courts will presume that a contract is legal. They will not, without more, presume an intention to violate the law. As we saw in *Murray Walter, Inc.*, above, a contract will not be deemed illegal if it is capable of an interpretation that will make it legal and allow the courts to uphold it. Even where a portion of a contract is clearly illegal, the courts will seek to sever the illegal portion and allow the legal portion to stand.

§ 8.1.3 Unconscionable Agreements

Just what is an unconscionable agreement? It is clear that the courts will not overturn an agreement that was entered into by the parties freely and with due deliberation. *Hydraulic Power Co. v. Pettebone-Cataract Paper Co.*, 198 A.D. 644, 191 N.Y.S. 12 (4th Dept. 1921), involved a dispute between Hydraulic Power, the owner of a hydroelectric system at Niagara Falls, and defendants Pettebone-Cataract and Cataract City Milling Co., who were owners of certain water rights. By means of a canal, Hydraulic Power tapped into the Niagara River above the falls and conveyed water to a basin located on the high bank of the river below the falls. From that basin, water was carried over the bank onto turbines near the river level in order to generate electric power.

Pettebone and Cataract had taken quantities of water from the basin pursuant to their water rights. The water was conducted to turbines upon their properties, thereby generating electric power for their own uses. The turbines of Pettebone and Cataract were located about 100 feet below the top of the high bank. After the water was run through those turbines, it was discharged on a sloping bank owned by Hydraulic Power, down which it flowed into the river. Hydraulic's predecessor in interest had entered into contracts with Pettebone and Cataract.

Prior to making those contracts, controversies had arisen between Hydraulic's predecessor and Pettebone and Cataract respecting past usage by Pettebone and Cataract of water from the basin. There was at least one litigation with respect to those controversies. After considerable negotiations and with the aid of able counsel the contracts were made. The contracts definitively fixed the quantity of water to which both Pettebone and Cataract were entitled in cubic feet per second, fixed payment for excess water used in the past, and detailed at great length the methods and agreements for the measuring of water to be used in the future. The contracts also stated that it was contemplated that at some time in the future, Hydraulic's predecessor or its successors would construct a tunnel, flume, or other conduit along the easterly bank of the Niagara River for the purpose of collecting water discharged by the mills operating on the top of said bank, and otherwise flowing over said bank. Such water would be accumulated and used in the development of power on the lower portion of the bank. The contracts granted the right to construct the tunnel and to collect and utilize the waste water from Pettebone's and Cataract's turbines. It provided that the tunnel should be at least 75 feet below the surface and that plans and sketches would be submitted to Pettebone and Cataract prior to the commencement of construction. Pettebone and Cataract could make written objection within 30 days. Upon the occurrence of differences as to the plans submitted, engineers of the parties would confer in attempting an adjustment. Should the engineers be unable to agree, a third engineer would be called in. If there were an inability to agree on the third engineer, then the third engineer would be the chief engineer of the Holyoke Water Power Company.

Nearly 15 years after the signing of the contract, Hydraulic Power as successor, served written notice on both Pettebone and Cataract of its proposal to commence construction of the tunnel and submitted a sketch in connection therewith. Both Pettebone and Cataract adopted a position antagonistic to that of Hydraulic Power. Each attempted to repudiate the contract and refused to join in any engineering conference. They also refused to select a third engineer. After waiting the period of time contemplated by the contract, Hydraulic Power called in the chief engineer of Holyoke Water Power Company and that engineer and Hydraulic Power's engineer passed on and approved the contemplated plan for the construction of the tunnel. Both Pettebone and Cataract urged that the contracts were unconscionable.

The courts will not enforce contracts that preserve the rights of only one party to the contract. The court here noted that although it did not dispute that principle, such is not the contractual situation present. The contracts were made with ample provision for settlement and adjustment of all disputes. Substantial benefits were acquired by all parties to the contract. Cataract and Pettebone cannot avail themselves of benefits and at the same time repudiate the burden of the contract.

The provisions of the contracts required Pettebone and Cataract to submit to the inconvenience of suspended operations for limited periods. They were coupled with corresponding provisions for payment of damages at figures that were not nominal. The court found that all matters were the subject of calm and deliberate adjustment. There was, therefore, not room for holding these contracts unconscionable or oppressive in any degree. Hydraulic Power prevailed.

In determining whether or not a contract is unconscionable, the courts will carefully examine the facts. High-pressure tactics and unequal bargaining positions have long been recognized to indicate unconscionability.

§ 8.1.4 Circumstances

Need actual injury have occurred in order to hold a contract invalid? *Black v. MTV Networks, Inc.*, 576 N.Y.S.2d 846 (1st Dept. 1991), involved a dispute between a consultant and MTV. The consultant, Black, had performed certain services for MTV and sought to be reimbursed for those services. Black had entered into a written agreement under which he was to provide services to MTV on retainer for two years. The contract provided for an automatic extension for another two years unless MTV notified Black of its intention not to renew.

MTV attempted to exercise its option not to renew, but mailed its letter to that effect too late. Black had in its possession a letter signed by MTV's former President of Network Operations confirming a two-year extension. MTV held to its termination and barred Black from its premises.

Black sued. MTV discovered that Black had secretly made gifts totaling thousands of dollars to MTV's Director of Personnel. He had also made an interest-free loan of $30,000 to another MTV employee. MTV argued that Black was prevented from any recovery. The contract had been fraudulently obtained, and Black had breached an implied covenant of good faith and fair dealing. Black conceded the payments and loan, but he argued that they were not intended to influence the behavior of MTV employees. Black claimed that the payments were made for personal reasons. He argued that the recipient of the gifts was not a decision maker but exercised purely administerial functions. The court found that the Director of Personnel was involved in Black's consulting work while she was receiving money from him. She was involved in the decision-making process and in a position to exert influence on his behalf. The court did not find that her loyalty was actually compromised.

A covenant of fair dealing and good faith is implicit in all contracts. Secret payments can erode an employees' duty of undivided loyalty to their employer. The court noted that Black's claim was that the gifts and loan were made to long-time friends in need. Be that as it may, the controlling factor is not the relationship between the giver and the recipients, but the fact that the transactions were kept hidden from the recipients' employer.

The vice lies in making the agreement without the knowledge of the master.

The court noted that regardless of intent, motive, illicit purpose, or pecuniary loss; secret payments improperly created interest for agents that were averse to that of their principals. The principals' complete knowledge and approval is required of any substantial advantage received by an agent from third persons. The concealment of such benefits from the principal, in and of itself, violated the covenant of fair dealing and good faith implied in every contract. It matters not that Black's payments were made to strangers or friends in need. The mere nondisclosure of Black's dealings with employees permitted the employer, MTV, to terminate the agreement. The court found against Black.

§ 8.1.5 Violation of Public Policy

When does public policy come into play, and what are its effects? The applicability of public policy ranges from circumstances mundane to those of the highest ethical import. *Hassett v. Trump Village Constr. Corp.*, 575 N.Y.S. 843 (1st Dept. 1991), involved an action for the death of an employee due to a scaffolding collapse. As general background information, it is important to note that any contributory negligence of Hassett would be irrelevant. The scaffolding law in New York was intended to shift the burden of insuring safety from the workers to the owners and general contractors. Liability was absolute.

The case involved the contracts among the various parties, the general contractor, owner, lessee, and subcontractor as well as the indemnification provisions found not only in their individual agreements, but under the common law. The general contractor was hoping to escape liability. The court noted that as against the subcontractor, a general contractor could only do so by showing as a matter of law that the subcontractor was actually responsible for the accident. The general contractor could not claim contractual indemnity because a promise to indemnify regardless of the general contractor's fault would be unenforceable as a matter of public policy. The court held against the general contractor's attempt to escape liability.

The public policy considerations in *Kalisch-Jarcho, Inc. v. New York*, 72 N.Y.2d 727, 536 N.Y.S.2d 419 (1988), were of a different nature. Kalisch-Jarcho was the plumbing contractor involved in the construction of a new broom depot and salt storage building for the Department of Sanitation. The contract contained a dispute resolution provision whereby the commissioner was to determine if disputed work was contract or extra work. The dispute concerned the excavation for and installation of concrete pads below certain underground fuel tanks. Kalisch-Jarcho was required to install the tanks and the concrete casing around them. However, Kalisch-Jarcho argued that the excavation for and installation of the concrete pads below the tanks was the responsibility of the general construction contractor. The commissioner ultimately determined that it was the intent of the contract documents that Kalisch-Jarcho was to perform that excavation and directed Kalisch-Jarcho to proceed with the disputed work.

Kalisch-Jarcho refused, arguing that it was not contractually obligated to perform the work and that the contractual provision violated public policy, which protects against collusive claims for extra work.

The history of that policy is instructive. The public policy argument was first advanced in the *Borough Constr. Co. v. City of New York*, 200 N.Y. 149 (1910). In *Borough,* the contractor sought recovery of extra work and materials in connection with the construction of a large sewer for the City of New York. The city engineer had ordered the contractor to lay a portion of the sewer in a more expensive cement. The contractor was also required to furnish an elevator to lower automobiles into the sewer, to clean up the sewer, and to illuminate it with candles for an inspection trip by city officials. Access was to be made more genial for those attending the opening day festivities. Borough had protested that the order was beyond the scope of the contract. It performed the work and then sought reimbursement. The court in *Borough* recognized that a contractor ordered to furnish services or materials covered by the contract, which the contractor believed were outside the contract scope, may, under protest, perform as directed, and recover. This is the case even if it turns out that the contractor was right and that the municipality had no right to call on it for such labor and materials. The general principle, then, is that the value of extra work may unquestionably be recovered by a contractor as a breach of contract damages.

However, the court weighed against this general principle, a potential for abuse. Municipal representatives and contractors could collusively saddle a municipality with uncontemplated liabilities. Therefore, the court imposed a limitation on the general principle. If a municipal representative's demand is flagrantly beyond the contract scope, Borough could not, even under protest, do the work and recover. Applying this test, the more expensive cement work, although not dreamed of by the parties and technically outside of the contract, could be sustained under a literal interpretation. Therefore, the contractor was justified in complying, and could recover. On the other hand, there could be no recovery for the other work. These demands were so preposterous that there could be no reasonable doubt that they exceeded the obligations of the contract. A refusal to comply with them would not work a breach of contract. Kalisch-Jarcho, reading *Borough*, was concerned that the direction by the commissioner would be clearly outside of the contract scope and violative of public policy. Even if Kalisch-Jarcho performed the work, the City could have argued it was outside the scope and justifiably refused to pay.

The court noted a material distinction between Kalisch-Jarcho's situation and that in *Borough.* The suit in *Borough* was commenced after completion of the work. The question was whether a contractor could recover for all the additional work. The Borough contract provided that the city engineer was to have final, binding authority on all questions relating to the work. It did not explicitly preserve the contractor's right to claim compensation for extra work. The question in *Borough* was whether Borough could sue for what it claimed was extra work. The essential question in Kalisch-Jarcho was not whether, but when. The contract in Kalisch-Jarcho contained a step-by-step procedure which, on one hand, preserved the city's right to have a multicontractor project proceed to conclusion without costly litigation interruptions and, on the other hand, preserved the contractor's right to ultimately seek judicial resolution of claims for addition compensation.

The court found a second material distinction. *Borough* clearly delineated what might be compensable, though technically outside the contract from what was preposterous, so as to suggest collusion, bad faith, or similar circumstances demanding protection of the municipality.

The court found that the excavation and installation of concrete pads below the underground fuel tank was shown on the contract drawings and indisputably part of the contract. There was no question of saddling the municipality with preposterous, unnecessary, or uncontemplated costs. The only dispute was whether the work was that of the general construction contractor or the plumbing contractor, Kalisch-Jarcho. The court found against Kalisch-Jarcho, holding that the excavation and installation of the concrete pads was not so preposterous or so manifestly beyond the limitations of the contract.

For spine-tingling illegality, there is nothing quite like the facts in *X.L.O. Concrete Corp. v. The Rivergate Corp.*, N.Y.L.J., Dec. 27, 1991, p. 22, col. 3 (N.Y. Co. 1991). The family activities here gives one pause. X.L.O. was a subcontractor for Rivergate, which was the general contractor with respect to a construction project. X.L.O. agreed to construct a concrete superstructure. After completing the project, X.L.O. demanded full payment of the contract balance due. Rivergate refused to pay. X.L.O. sued for breach of contract. However, in the court's eyes, this was not a simple breach of contract. The claim was set against the backdrop of the criminal monopolization of the concrete construction industry.

What appears to be on its face an ordinary action for breach of contract—you agreed to pay, and you did not— is elevated to notoriety because of particular family affairs. The family here being the Cosa Nostra.

The breadth of the court's decision is astonishing. For ease of understanding, we will move from the particular to the general.

X.L.O. was a New York corporation. Its principal was James Costigan. Costigan was a principal witness in three federal criminal trials against various organized crime figures. Costigan had pleaded guilty to a misdemeanor with regard to payment of a bribe to Ralph Scopo, President and Business Manager of the District Council of Cement and Concrete Workers, Labors International Union of North America. Scopo was a member of an organized crime family. Costigan was granted immunity from further criminal charges in exchange for his testimony.

Rivergate Corporation was a New York corporation. Its principal was Donald Zucker. Zucker was not a party in the federal actions. In defending against X.L.O.'s claim, Rivergate contended that the contract should have no effect since it was illegal and void. Every contract whereby a monopoly in any business or trade is established or maintained would violate the law and be void and unenforceable. Rivergate argued that X.L.O. was seeking to enforce a void contract resulting from illegal price fixing. It argued that the illegal price fixing and market allocation was run by organized crime in which X.L.O. was a participant. The crux of Rivergate's argument, as determined by the court, was that the contract sued upon by X.L.O. should not be enforced because it was obtained through an illegal conspiracy.

X.L.O. contended that the contract was not void and is enforceable. The court noted that the evidence did not reveal that the price agreed to between X.L.O. was unreasonable or fixed. The evidence did not reveal that the contract was completely collateral and independent, even though the agreed price was not shown to have been determined or controlled by those persons involved in an illegal conspiracy.

These are the particulars as the court observed them. The court next dealt with the background of the criminal activity. I am indebted to the court for the genealogy it provided in its decision.

The illegal conspiracy was known as the "commission" of the Cosa Nostra. Cosa Nostra was a nationwide criminal society operating through local groups known as "families." The commission oversaw the activities of five families, the Genovese, Gambino, Colombo, Lucchese, and Bonanno families, in New York City with affiliated families in other cities. Its general purpose was to regulate the various family ventures and resolve disputes among the families. The bosses of each family were, respectively, Anthony Salerno, Paul Castellano, Carmine Persico, and Anthony Corallo. Due to internal instability, the Bonanno family was denied a seat on the commission during the period in question. The court noted that various members of the families were convicted of conspiracy or labor bribery violations.

What concerns us here is the extortion and labor bribery scheme known as the "club." This venture involved the four New York City families. Through the commission and Scopo's control of the district council, an arrangement was made with seven concrete construction companies working in New York City. The rules of the club provided that only construction companies approved by the commission would be permitted to take concrete construction jobs worth more than $2 million in New York City. Any contractor taking a job worth more than $2 million was required to pay the commission 2 percent of the construction contract price. The commission would approve which construction company in the club would get the job. The bids would be rigged so that the designated company submitted the lowest bid, circumventing any competitive bidding. The commission would guarantee labor peace in exchange for compliance with the rules of the club. The commission would enforce compliance by threatened or actual labor unrest or physical harm. In extremes, the commission could drive a construction company out of the concrete business. The court noted that according to federal law enforcement authorities, seven concrete companies were participants in the scheme. X.L.O. was one of the participants. Another participant was forced out of the concrete business because it would no longer make payments to the club.

In a federal action involving the club's scheme, X.L.O. had been described as a victim of extortion or attempted extortion by the commission. Costigan had testified that the commission used the threat of labor unrest, the disruption of concrete supplies, and its reputation for violence to enforce the rules of the club.

Costigan did not argue that the findings of the federal court were untrue. He maintained that Rivergate and Zucker were well aware of the club and were working along with it. X.L.O. alleged that Rivergate profited from the dilemma faced by X.L.O.

Zucker was aware that he could not, as a general contractor, obtain competitive bids for the work because of the operation of the club. Zucker had used X.L.O. before and understood that X.L.O. alone would negotiate for the work.

For its part, X.L.O. argued that it had resisted the threats of the commission until it was the last concrete superstructure doing business in New York City, which was not a member of the club. The pressure became too great. Scopo alleged he could no longer control the situation. At that point X.L.O. joined the club. X.L.O. acknowledged that it was required to pay the 2 percent fee on each contract it was awarded. X.L.O. and Rivergate negotiated a contract price of $16,300,000. X.L.O. paid the commission $320,000. X.L.O. contended that when the commission learned of the $16,300,000 figure, it caused a major uproar among the four families because X.L.O. was not supposed to get projects of over $15 million.

At this juncture, Costigan was approached by Scopo and advised to walk away from the job. Costigan told Scopo that he would not relinquish the agreement with Rivergate because the club had not gotten X.L.O. any work for over 18 months. At a subsequent meeting, Scopo informed Costigan that he had spoken up for Costigan at a club meeting, and the four families decided that X.L.O. could keep the project without having to pay any penalty. Costigan admitted that in addition to the 2 percent club fee, he paid Scopo $50,000 as a gift for speaking on X.L.O.'s behalf.

The court noted that there was no indication that Rivergate was coerced into negotiating with X.L.O., nor that Rivergate was prevented from soliciting bids from concrete contractors outside of New York City. The record did indicate, however, X.L.O.'s status as a member of the club and Scopo's support of X.L.O. These were major factors in the allocation and retention of the project by X.L.O. The allocation was possible because the commission had monopolized the concrete construction industry in New York City by threats, intimidation, and control over unionized labor.

The court noted that X.L.O. sought to recover under the contract. The court also highlighted that, but for the illegal conspiracy, the contract would not have come into being. It noted that the contract was the result of job allocation made by the commission and that the application was an integral part of the illegal conspiracy among the seven concrete companies. While the seven concrete companies may have been victims and the contractors may have been beneficiaries, the illegal conspiracy operated under the auspices of organized crime. The contract came about as the result of the involvement of organized crime, a corrupt president of a local union, and the passive or active participation of contractors and subcontractors, including X.L.O. and Rivergate.

However, the court emphasized that X.L.O. cannot expect a court to lend its support to illegal actions by way of enforcing the contract. Persons must understand that when they enter into contracts of this nature they place themselves outside of the protection of the law. The better persons understand this, the less inclined they will be to enter into such contracts.

The court noted that both parties could not dispute that they were aware of the illegal activity and were either active or passive participants before and after the execution of their contract. Although X.L.O. argued that the contract was independent of the conspiracy, the court found that the record indicated otherwise. Costigan, without reservation, admitted that X.L.O. was allocated the project by the commission as a club member with the intervention of Scopo. But for the illegal conspiracy, X.L.O.'s club membership, and the gift to Scopo, the contract would not have been entered into and retained. The court found that the conspiracy was illegal and in violation of the law. The violation was integrally related to the contract sued upon. Both X.L.O. and Rivergate knew that the execution of their contract was a foregone conclusion. The project had been allocated to X.L.O., and Rivergate understood beforehand that it was to negotiate the contract with X.L.O. The contract was an integral part of the illegal conspiracy. The court noted that X.L.O. characterized itself as a victim of illegal activity. The court emphasized, however, that its aiding and enforcing the contract would be tantamount to assisting in the realization of the fruits of an agreement tainted with illegality.

The aid of the court was denied, not to benefit Rivergate, but because public policy demanded that enforcement must be denied without regard to the interest of the individual parties. The court noted that whatever contamination reaches, it destroys. The court found against X.L.O. and denied enforcement of the contract.

§ 8.1.6 Waiver of Statutory Right

Waiver of a right provided by a statute is clearly indicated in yet another scaffolding law situation. In *Horning v. Gore*, 87 A.D.2d 34, 451 N.Y.S.2d 848 (3rd Dept. 1982), Horning was a contractor building a single-family dwelling for Gore. Horning sustained injuries when a scaffold, which he had designed and erected, collapsed causing him to fall 12 feet to the ground. In general, the legislature has seen fit to afford certain protection to workmen in the building construction, repair, and demolition industries. An owner would be held liable for scaffolding and other protective devices for the use of workmen. However, here, Horning was the signatory to a contract.

The contract provided that Horning was to perform the work in a good and workmanlike manner and to provide good, proper, and sufficient materials, labor, and scaffolding necessary for the construction. The contract continued that Gore would not in any manner be answerable or liable for any of the materials or implements used or employed in the erection or furnishing of the house.

The court found that by signing the contract, which excluded any manner of liability for implements used in the erection of the work, Horning had expressly waived the protection afforded him under the New York Labor Law. This would not be the case had Horning been an ordinary workman. Here, Horning, however was the actual signatory of the contract. The court found no reason why Horning should not be bound by its terms. The court found for Gore.

8.2 LEGALITY OF CONTRACTS

§ 8.2.1 Express Legislative Intent

As we have seen, a contract that somehow violates the law will be deemed by the courts to be illegal and void. Just how do the courts determine that a particular contract is violative of the law? The court will often examine and interpret the intent of the legislature.

Stephen Peabody, Jr. & Co. v. Travelers' Ins. Co., 240 N.Y. 511 (1925), reh'g denied 241 N.Y. 542 (1925), involved a dispute regarding workmen's compensation insurance. Peabody was an insurance broker. Peabody had obtained certain workmen's compensation insurance for the New York Dock Company. Peabody argued that it had reached an agreement with the Travelers' Insurance Co. whereby Travelers' would not increase the insurance rates charged to New York Dock, Peabody's customer. The workmen's compensation law had required that every insurance company supplying workmen's compensation insurance file with the Superintendent of Insurance its classification of risks and premiums, together with base rates and schedules. The Superintendent of Insurance would need to approve the same before they could take effect. The Superintendent of Insurance could withdraw his approval of any rate or schedule if, in his opinion, the rate or schedule was inadequate to provide necessary reserves.

Travelers' was required to advise the Superintendent of Insurance of its classifications of risks and premiums, its basis rates and schedules. Travelers' had agreed that in order to permit New York Dock to know in advance that its premium rates would not exceed a certain amount, Travelers' would write a policy at rates not to exceed a certain amount irrespective that premium rates might be fixed at a higher rate by the Superintendent of Insurance.

The crux of Peabody's dispute was that the rates charged by Travelers' should not be higher than a certain fixed and definite amount irrespective of the actions of the Superintendent of Insurance. Peabody claimed that Travelers' broke its agreement not to increase insurance rates and injured Peabody's business and its relationship with New York Dock.

The court found that such an agreement would be clearly against the spirit if not the letter of the law. The insurance rates for workmen's compensation premiums, the basis rates, and schedules were to be fixed by the state insurance department. They were not be fixed by Travelers'. It was the duty of the Superintendent of Insurance to see that the rates and premiums were adequate to protect the employees insured and to preserve and provide the necessary reserve funds. The court found that a contract that was alleged to disregard any increase in the basis rate was against public policy and void. An insurance broker such as Peabody is supposed to know the law as well as Travelers', the insurance company. Travelers' had no right under the insurance laws to fix a rate of premium not acceptable by the insurance department. Peabody, as broker, would have no right to procure from it a contract to do so.

Such a contract would be illegal. An insurance company having made such a contract could not be held to it and would violate no right of the insurance broker by repudiating it. The law favors obedience to its statutes, rather than the keeping of contracts in violation of them. The law will not recognize as a breach, that which is in obedience to a statute. Nor will the courts enforce a contract in violation of a statute. The court found for Travelers'.

§ 8.2.2 Purpose of Statute

As a contractor, you may be engaged by a home owner to perform additions or renovations. The question of local licensing will arise in many counties and municipalities. The licensing statutes for home improvement contractors are strictly construed. Courts will look at the purpose of the statute and act accordingly. *Estates Roofing Co., Inc. v. Homeowner*, 121 Misc.2d 279, 464 N.Y.S.2d 951 (Civil Ct. Queens Co. 1983), involved a home improvement contract for certain roof repairs with a homeowner who was an attorney. The homeowner paid about one quarter of the contract amount. He did not contend that the repairs were completed in anything less than a workmanlike fashion. He only argued that Estates Roofing was unlicensed at the time the repairs were made. The court found that the licensing legislation was enacted by the City Council for the safety and protection of the public. While there was no dispute as to whether the work was done by Estates Roofing or whether it was satisfactorily completed, the court found for the homeowner. The court stated that the statute was enacted to protect the public. Although the result appears to be unjust, "the law is the law."

Zimmett v. Professional Acoustics, Ltd., 103 Misc.2d 971, 431 N.Y.S.2d 243 (App. T. 1980), was a dispute between the owner of a cooperative apartment and his contractor in connection with the performance of certain structural repairs.

The court noted the general rule that a contract, although in violation of a statute, will not ordinarily be declared invalid where it was not the intent of the legislature to make it void. If a statute violated does not provide that contracts involving its coverage shall be void, then the legislative intent must be ascertained from the statute as a whole. The court noted that licensing statutes are designed to protect the public against fraud. It had been held that failure to obtain the required license rendered the contract unenforceable.

Professional Acoustics had argued that it was not required to be licensed because it was a subcontractor to the owner's architect. The court found unpersuasive the fact that Professional Acoustics was performing the home improvements under the supervision and control of the architect. It found that Professional Acoustics had attempted to frustrate the intent and purpose of the statute and to contradict its express provisions. The statute had made no distinction between general contractors and subcontractors. Its provisions were specifically intended to protect a homeowner against abuse and fraud practiced by persons engaged in home improvement. The court found that if subcontractors were not required to comply with the provisions of the statute, claiming exemption by reason of an architect's or general contractor's license, the statute would be emasculated. Zimmett was not required to pay the contract balance.

But what if the statutes are in conflict? *Lettmann v. Patterson*, N.Y.L.J., May 3, 1991, p. 30, col. 3 (Kings Co. 1991), involved a contract wherein Lettmann retained the services of Patterson, an unlicensed contractor. Lettmann, an experienced real estate broker, was rehabilitating a multiple-dwelling building for profit. Patterson was to install floors, a roof, staircases, and gaslines as well as providing the materials for the construction. The work performed by Patterson clearly fell under the home improvement licensing statutes. Furthermore, there was no dispute that Patterson was an unlicensed contractor. However, the court looked at yet another statute, which defined an owner as any homeowner, cooperative shareholder, residential tenant, or person who purchases a custom home.

The court looked to the legislative history. It noted that the licensing scheme for home improvement businesses was clearly designed to protect the homeowner against fraud. However, the court found that an experienced real estate broker, seeking to rehabilitate a multiple dwelling for profit is not among the class of consumers or owners meant to be protected. Lettmann, in a business venture, would be better able to protect herself in dealing with an unlicensed contractor than would the owner of a home who might reside in the subject property. The court held against Lettmann.

§ 8.2.3 Partial Illegality

Can a portion of a contract be illegal? And, if a portion is illegal, can any part of the contract be saved? Remember, illegality taints all within its reach, but where a contract may be severed, the illegal portion of the contract may be rejected and the valid portion retained. But can the contract be severed?

American Store Equipment & Constr. Corp. v. Jack Dempsey's Punch Bowl, Inc., 174 Misc. 436, 21 N.Y.S.2d 117 (N.Y. Co. 1939), aff'd 258 A.D. 794, 16 N.Y.S.2d 702 (1st Dept. 1939), aff'd 283 N.Y. 601 (1940), involved a contract to aid in and supervise the construction of Jack Dempsey's restaurant. As a portion of its contractual work, American performed certain architectural services, including the drawing, preparing, and furnishing of plans and specifications. American clearly had no license as an architect and could not be licensed. The law regarding the licensing of architects was promulgated to safeguard life, health, and property. Included in the practice of architecture was any professional service such as consultation, planning, and design, including aesthetic and structural design, and responsible supervision of construction in connection with private and public buildings. American's contract required the preparation of plans and the supervision of construction work that is usually the function of an architect. American could not show compliance with the architectural licensing statutes.

American contended, however, that even if it did perform certain architectural services, it can recover for the portions not regarding those services. It argued that the architectural services, if any, amounted to only about five or ten percent of the total services undertaken. The court found, however, that there was no means of segregating the good from the bad portions of the contract. The contract was entire and indivisible: to plan, construct, and furnish a complete unit. If American had sold interior furnishing and decorations, the contract could have been separated, at least to the extent permitting recovery for merchandise sold. This contract, however, had involved only services, ideas, and supervision. It cannot be separated into separate classes legal and illegal.

The moral of the story is to have a severability provision in your contract. A severability provision would direct that should some portion of the contract be deemed illegal, it can be severed with the remainder of the contract continuing in effect. The court in *American* considered that if the contract is contrary to public policy, the whole contract fails. This contract arose out of and was indissolubly linked with an illegal bargain.

The court noted that if it were to sustain the legality of the balance of the contract, it would foster a widespread disregard of the licensing statutes. It noted that it would be easy for a construction contractor to thwart the purposes of the licensing statutes by merely providing in his contract that architectural services will be given gratis, so long as the contractor were to be awarded the contract itself. The court would not countenance the behavior and found for Jack Dempsey.

§ 8.2.4 Contracts Involving Criminal Matters

S.T. Grand, Inc. v. New York, 32 N.Y.2d 300, 344 N.Y.S.2d 938 (1973), involved a dispute over the unpaid balance due on a reservoir cleaning contract.

There was no competitive bidding since James Marcus, the commissioner of water supply, gas, and electricity let the contract pursuant to a "public emergency" exception to the general bidding requirements. The entire cleaning had been performed. Subsequently, Grand and its president were convicted in federal court of conspiracy with intent to violate New York State bribery laws. The conviction was based upon a series of events culminating in an agreement by Grand to pay a kickback to Marcus in return for Marcus' award of the cleaning contract to Grand.

When Grand sought its unpaid balance, the City countered that the contract was illegal and requested that it receive the amount previously paid to Grand. The court first examined whether or not the criminal conviction of Grand for bribery conclusively establishes the illegality of the "emergency" cleaning contract. The court noted that Grand was convicted of conspiracy to use interstate facilities with the intent to violate the bribery laws. The facts upon which that conviction rested—the bribery of Marcus—were identical with the facts asserted in this case to prove the illegality of the "emergency" cleaning contract. The court noted that Grand was afforded a full and fair opportunity to contest the bribery issue at its trial. There are rigorous safeguards imposed to insure against unjust conviction in a criminal action. In light of that, the court found that the criminal conviction was conclusive proof of Grand's bribery of Marcus. The illegality of the "emergency" cleaning project was established.

The court then proceeded to the general rule: where work is done pursuant to an illegal municipal contract, no recovery may be had by the contractor. The municipality can recover from the contractor all amounts paid under the illegal contract. The court noted that the reason for this harsh rule is to deter violations of the bidding statutes. Although the nature of the wrong is such that it is not easily discovered, the courts will make it clear that they will decline to aid the fraudulent contractor who seeks recovery. The court noted that the illegality went to the very origin of this contract.

In such a circumstance, the general rule of complete forfeiture is applicable. Although the result is harsh, it is necessary where bribery of municipal officials is confirmed. The court found for the City and ordered that Grand return all monies previously paid on the illegal contract.

§ 8.2.5 Public Contracts

What if the courts are obviously unhappy with illegality in connection with a public contract? How do the courts view the practice of employing someone to procure a government contract? Obviously, if corrupt means are employed, the agreement would be against public policy and void.

Dunham v. Hastings Pavement Co., 56 A.D. 244, 67 N.Y.S. 632 (1st Dept. 1900), reargument denied 57 A.D. 426, 68 N.Y.S. 221 (1st Dept. 1901), involved an agency contract between Dunham and Hastings. Dunham was to be the sole agent for Hastings in the City of New York. Hastings manufactured paving blocks known as compressed asphalt paving blocks. The City of New York had used only sheet asphalt for paving. Dunham organized a company and arranged to lay a sample pavement of asphalt block in the city. He reported to Hastings that the city authorities would require paving to be laid for five years before allowing further bidding using asphalt block pavement. At the end of the five years, the exclusive agency contract was made. The contract recited that Hastings was desirous of having its paving blocks used in New York City. Since Dunham had offered to undertake and perform the labor of securing the consent of the City, he was to have exclusive right during a limited period of introducing and securing the use of the paving blocks. The contract provided for compensation to Dunham based on the number of square yards laid under each contract secured.

Dunham had subsequently secured for Hastings Pavement the right to bid on various other paving contracts. The right was based on the passing of resolutions by Board of Aldermen for the paving of unpaved streets and the approval by the Mayor. It also entailed interviews with city officials who were expert in street paving, visits with them to other cities where the asphalt paving blocks were in use and various interviews with the mayor, commissioner, and members of the Board of Aldermen.

Consequently, a large amount of the asphalt paving blocks were laid in the City. Hastings Pavement did not pay the commissions, and Dunham initiated suit.

The court noted that a contract is not necessarily void because the enforcement of rights under that contract may violate a statute. The general rule is otherwise, however, where the public is affected by violation of a particular statute. Where the enforcement of rights arising under a contract is opposed to good morals or sound public policy, the court will refuse to aid parties. The parties will be left as the court found them. The court noted that the contract, on its face, was not tainted with illegality. If the purpose of the contract is to be deemed illegal, it is because use of personal solicitation, the exercise of personal influence, or the execution of improper or corrupt methods to procure a legislative or official action.

If a contract contemplated legal service, and that alone, it would not be rendered illegal by the fact Dunham performed illegal acts in its performance. The question was, and continues to be, does the contract in fact require performance of illegal services? If it does not, then it is valid and can be enforced. One or several illegal acts performed under a contract may not render that contract void. Do the parties to the contract contemplate performance of illegal acts?

The court would not stop to separate legal from illegal acts in order to uphold a contract in part. If the contract contemplated but one illegal act, it would vitiate the entire contract.

The court examined Dunham's acts. Clearly, the contract would require him to see the Mayor and other administrative officials. It is not apparent how he could have proceeded to discuss the matter with them, except by visiting them at their offices. That was the place where they transacted public business and the place where Dunham had a right to go and where the performance of the contract required that he should go. Dunham had the right to present to these officers the merit of the pavement, which he desired to have introduced, and it was proper, if he could so induce, to have them inspect not only the pavement laid in New York City but laid elsewhere. If Dunham did no more than this with these officials, the court would be unable to see how his acts were improper.

Dunham's dealings with the Aldermen stand on a different basis and would receive more rigid scrutiny. If all Dunham did was recommend the merit of the pavement and convince them to examine the pavement that had been installed for five years, there would exist no basis for criticism.

The court noted that the difficulty with the illegality argument lies with the liability and opportunity for abuse. The court found that the contract was not void on its face.

A final case gives credence to the thought that the poor will be always with us along with the rich, the corrupt, and the incompetent. *Chard v. Ryan-Parker Const. Co.*, 182 A.D. 455, 169 N.Y.S. 622 (1st Dept. 1918), involved a contract between Cornelius J. Sullivan, the cousin of Timothy D. Sullivan, a former state senator and member of Congress, and the Ryan-Parker Construction Co. Here, the court found that the illegality threshold had been exceeded.

At issue was a simple letter contract that Cornelius Sullivan would render undefined services to Ryan-Parker and that in the event that a contract for the construction of the Manhattan Bridge was awarded to Ryan-Parker, Sullivan would receive one half of the net profits for his services, "in a managing or advisory capacity." The court noted that Sullivan was engaged in the life insurance business. He was also connected with Sullivan Advertising Company. Sullivan was not an engineer and never had anything to do with bridge construction. He was not a lawyer and had never studied law. However, the court noted that Sullivan and former Senator Sullivan did have certain interests in common, and it was their custom to meet at the rooms of the Timothy D. Sullivan Association practically every day when they were both in New York City.

Sullivan's contention relied on certain basic facts. In 1905, the City of New York had received bids for the construction of the Manhattan Bridge. The lowest bid was somewhat in excess of $7 million. That amount was the result of an agreement among the bidders and the courts enjoined the letting of the contract.

Sullivan claimed that the other bidders were still in agreement to keep the price high. Sullivan's idea was to put in a bid, which would result in a substantial profit, yet would be sufficiently low to make obtaining the contract probable. He discussed this plan with Patrick Ryan, President of Ryan-Parker. Sullivan worked with Ryan in elaborating the project and determining the figures. Sullivan testified that by studying the schedules employed on the previous bidding and considering the estimates made by Ryan and his engineer, he was able to advise that Ryan-Parker place a bid at approximately $6.5 million. This was substantially the bid of Ryan-Parker, and the contract was awarded to it.

The court noted that Sullivan did not explain how he was able to make a bid that would cover costs and profit without any knowledge of bridge or other construction. The court also noted that Ryan-Parker's engineer recovered a judgment of $40,000 for bringing this matter to the attention of Ryan-Parker and preparing the bid.

Sullivan claimed that he conducted negotiations for materials, dealt with the surety and comptroller regarding a bond, and interviewed various concerns at the request of Ryan. He claimed he negotiated with the company who supplied the structural steel and that he was involved with the contract for cables and for miscellaneous construction work. In addition, Sullivan claimed that for seven months he devoted all of his time to forwarding the project and carrying out every instruction given to him by Ryan.

Ryan-Parker argued that Sullivan had never rendered any services of value whatsoever. In defense, Ryan-Parker claimed that the intention and purpose of the letter agreement was to procure political influence and obtain the award of the bridge contract. In addition, once the contract was obtained, the influence of Sullivan was to render it secure against attack. Ryan-Parker argued that the letter agreement was void as against public policy.

The court noted the extraordinary nature of this letter contract. A corporation has agreed, instead of paying dividends to its stockholders, to divide its property running into hundreds of thousands of dollars with a person who has rendered no specific service whatever. Yet, the court also noted that the contract was not corrupt or illegal on its face. Contracts for the sale of political influence are not apt to so provide in their terms or express obligations, the intent to employ corruption or improper recourse to public officials. The court found that where the contract contemplates official action as a condition precedent for payment of compensation and where a huge sum is paid for unspecified services vaguely and generally referred to, the court must look to surrounding circumstances. The court considered the situation of the parties, their negotiations, and the subject matter of the agreement.

The law cannot lend its aid to enforce a contract that tends to corrupt or contaminate the integrity of the social or political institutions. The question turns not upon whether improper influences were contemplated or used, but upon corrupting tendency of the agreement. The court noted that the recovery of compensation for services actually rendered may be allowed, even under a contract that tends to secure corrupt actions or improper resort to public officials, in the absence that such was the intention of the contract. However, where there is evidence that warrants a finding that the intention of a contract, entirely innocent on its face, was to secure official action by means of political considerations or other improper means, the law cannot aid in recovery; not even if the services actually rendered were legitimate.

The court noted the inference that the contract contemplated obtaining the political influence of former Senator Sullivan to assist Ryan-Parker in getting the bridge contract and carrying it out without subsequent hinderance.

Ryan testified that just before bids were opened, he had a conversation with Sullivan and told him he wanted to see former Senator Sullivan. Sullivan made the appointment and Ryan met with former Senator Sullivan at the clubhouse of the Sullivan Association. Former Senator Sullivan told Ryan that he had passed the resolution in the Senate, making it possible to construct the Manhattan Bridge. The senator continued that he could furnish a bond and a $75,000 certified check. Ryan never received the bond or the check from the senator. Two employees of Ryan went down on the bond. Ryan's lawyer had turned over to these straw bondsmen a quantity of gold mining stock which the lawyer had held as trustee in order to qualify the bondsmen. Ryan knew that the bond would be attacked and could not stand an investigation.

The court noted evidence upon which a jury might find that Ryan was purchasing political influence. Sullivan testified that when he brought Ryan to meet the senator, Ryan asked the senator, "Senator, how far will you go for Con?" The senator looked him in the eye and said, "Mr. Ryan, I will go as far as I can. Is that satisfactory?" Ryan replied, "That is satisfactory to me."

The court noted that given the circumstances of the situation and the relationship of the parties, the jury could have inferred that this was a plain promise on the part of the senator to put his political influence and backing behind the project in which Cornelius Sullivan was interested.

The court considered the following matters: the vagueness of the services as stated in the letter contract, the fact that Sullivan's compensation depended upon the contract being awarded, the fact that Sullivan was not in any way connected with the bridge building business, and the very large sum claimed for the services. In spite of the conflicting testimony, an inference could be drawn as to the real purpose and intention of this contract. The jury should have been allowed to determine if there was illegal intent in the letter contract and if it was made to secure the political services and influences of Senator Sullivan in securing the award of the bridge contract.

CITATIONS

INDEX

Printed in the United States
122885LV00007B/3-4/A

9 780471 574149